Shavings for Breakfast

The history of The Morris Furniture Company

Glasgow

By

Charles E. Mac Kay

1884 - 1975

Ce qui vous ont à votre coeur

For Julie McCulloch

Published and graphics by

A. MacKay

G12 0DY

Printed by Clydeside Press, 37 High Street, Glasgow, G1 1LX

ISBN: 978-0-9573443-1-0 Editor: Iain C. MacKay
First published: 2013

Panelling the Queen Mary

A 44000lb bomb NBR

Making history making furniture

"It is not often that one has the privilege of spending a lifetime working with materials that have taken three or four hundred years to mature and also with a body of people whose dedication and sincerity to their craft is paramount. It was my privilege to work with such people; on good jobs, using good materials, and good equipment, during a period of over forty years. It is only fortunate that we managed to do proportionately more good things than bad; otherwise we would never have been allowed to survive. I cannot adequately express the excitement aroused by the use of a new machine, the start of a new job, the discovery of a new material, or a new way of doing something. It was all creative; even the contradiction and the criticism was constructive, everything contributing to constant improvements in our work. There may, in this book, appear to be an excess of name-dropping, for which I make no apology. Life after all is made up of people and it is people and their qualities that help to shape the colour of one's life" (Adapted)

Neil Morris

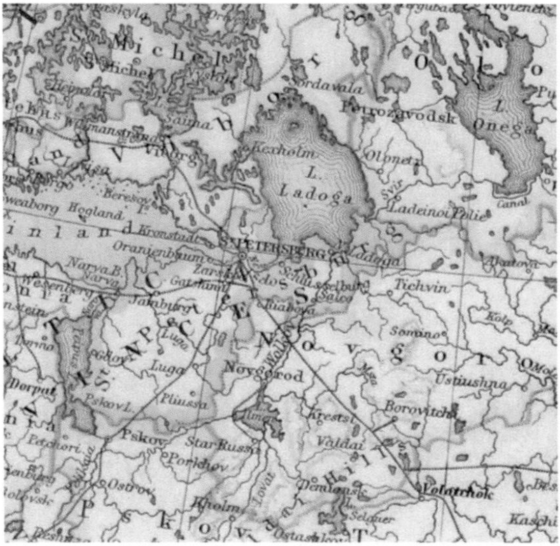

Saint Petersburg at the time of the Tsar. The city dominated Imperial Russia as an entrepôt and the centre for plywood manufacture

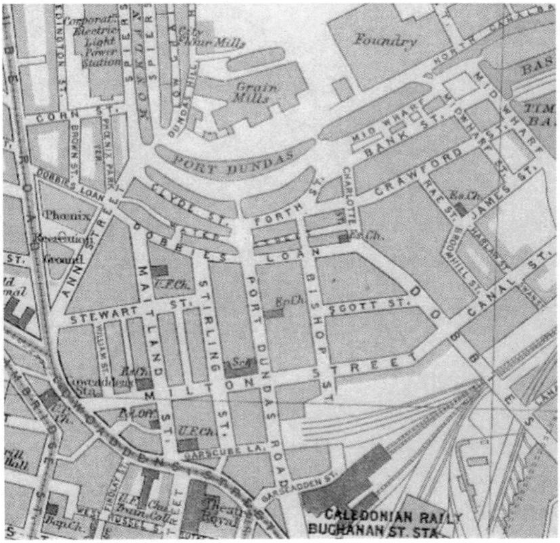

Milton Street in 1914 Glasgow. The factory was at the corner of Dobbie's Loan and Milton Street

The Canadian Pacific liner Empress of France. Morris extensively outfitted her as the S.S. Alsatian for the Alan Line. Her panelling is reported to be in a house at Glasgow's Bellahouston Park.

Milton Street, in Cowcaddens, producing cabinetwork for London County Council. The youngest employee appears to be around fourteen

Contents

The Scottish Aviation Liberator civilian conversion flying with Morris nose and wooden internal fittings

Gleneagles Hotel, the Riviera of the Highlands and the first major hotel contract for the Morris Company, furnished around 1922 – 23

Compilers Note

I first met Neil B. Morris around 1987 when he had asked me to help him with his researches. I was doing a book on the Beardmore Company at the time – Beardmore Aviation. Morris Furniture had manufactured target drones for the Royal Navy for use on H.M.S. Argus and I needed confirmation that such a contract existed. It took him two years to write back, but he confirmed work on the drones, Bennie Railplane and the Beardmore Taxi at Anniesland. He wanted to do a further book on the history of his family firm, H. Morris & Co. Ltd. The book was to fill the gaps in the original "Shavings for Breakfast." Knowing of my research experience, he asked me to do the war years and the helicopter, he would cover the pre-war years and the post war years and I would research all the details and include the findings in a text. Out of the blue he asked me if I knew about Rouken Glen Park. I said I knew very little about the place. He asked me to quickly research the history of the park and prepare a text. This was Christmas 1994. I dropped all my other research and attended to his needs. I did not know that he was seriously ill and was devastated when he passed away in 1995.

The firm had moved from Cowcaddens to Drakemire Drive near Castlemilk and, with the move, the company records were deposited at the University of Glasgow. I was asked to catalogue the massive collection for the firm. Neil Morris and I had worked on the wartime chapters and the helicopter and discussed the salient points agreeing the facts, but life moved on and the project was put on hold. I continued my interest in the collection and eventually prepared a script and I spent many summers going over the collection at the university. I noted papers and illustrations and where everything was. The accessibility of the PC and word processor made the task a lot simpler. When I initially wrote for Neil Morris it was all handwriting and cut and paste, literally, scissors and glue, but the PC made it a lot easier.

In researching the book I saw how fashions changed and how the markets reflected the needs of the customer whether it be a private house, a shipping company or the military.

The archive is a very important part of any company because it shows clearly how the company grows and develops over the years and how production policy was arrived at. Sadly, with the growth of new technology, the idea of an archive seems now to be obsolete with electronic technology replacing the paper driven office. Somehow there has to be a way of keeping a permanent record, which could be accessed to later generations showing clearly how a firm grows and develops from the past into the future.

Charles E. Mac Kay, Kelvindale, Glasgow

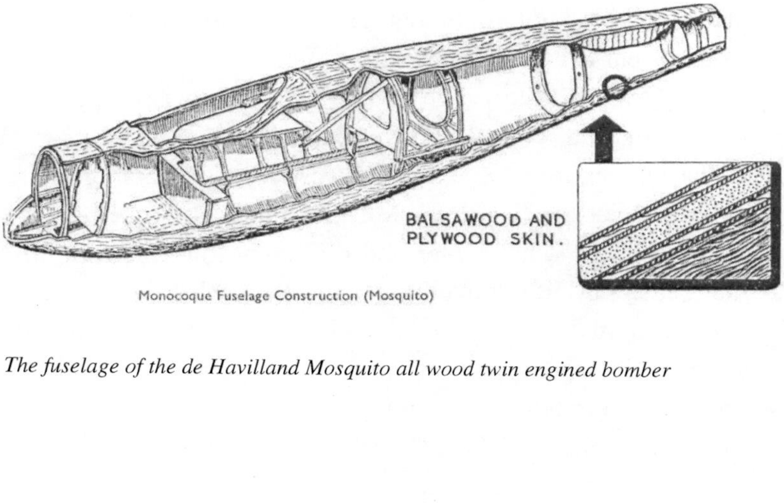

The fuselage of the de Havilland Mosquito all wood twin engined bomber

Acknowledgements

This project has taken many years to come to fruition and would not have been possible to write without the help and expertise of the following organisations and individuals.

The late Neil B. Morris was a fount of detailed information. I am grateful for his guidance and friendship during the formative years of the project. He proof read the wartime chapters of the company and the chapter on the helicopter and others, he pointed out its discrepancies, added accurate information and, even late on in life, he was still frustrated by lack of progress on the Flying Jeep project.

The staffs of the University of Glasgow Archives were a great help with their friendly advice and assistance. Emrys Inker of Weir's Pumps ably assisted me in my search for information and alternative sources for helicopter development as did Peter Batten formerly of Westland Helicopters. Mr W. Carter helped by explaining the role of Barnes N. Wallis in British aerial bomb development and the role of Vickers and Vickers Armstrongs in the production of armaments before and during the Second World War. The American interest was ably met by Mr John Godden. The late H.J. Woodend of the Ministry of Defence Pattern Room at Nottingham was always ready to provide accurate information on rifle production. The librarian of VSEL, Mr W. Wallace, also assisted in the identification of Vickers built commercial vessels that Morris furnished, through hull numbers; he accurately pointed out to me the difference in the Bermuda vessels.

The librarians and staff of the Mitchell Library were a tower of strength during the researches of this book; this is where the bibliography was to hand. As the research of the book was underway the Mitchell Library went through a long series of painful restructuring plans and alterations. The Librarian at the Royal Aeronautical Society supplied information from privately published periodicals that included lectures by Neil Morris on the production of the wooden helicopter blade for Cierva and Bristol Helicopters and the use of autoclaves. The Archivist at the University of Strathclyde provided information on the Royal Technical College's work on the wind tunnel and helicopter blades carried out pre 1939 - 1945. The Scottish National Gallery for Modern Art for help in relation to the works of Benno Schotz. I am grateful to them all for their assistance and kindness and many thanks to Pauline Megson for the exchange of information during the early life of the project.

The family of the late Flight Lieutenant Władysław J. Świątecki, Polish Air Force, inventor of the heavy bomb slip device, recently supplied information that confirmed dates on the wartime activities of the company in relation to heavy bomb work and the Royal Commission on Awards to Inventors.

Julie McCulloch gave advice on how to restart the text of the book and volunteered to vet the manuscript and type up any forthcoming amendments. She encouraged me to work out the length of the text and the shape of the book itself. This was as Neil Morris would have wished it to be. To my family and friends and others I am deeply grateful for all their encouragement, support and enthusiasm during the course of this project.

Charles E. Mac Kay Saint Andrews July 2012

Notes on Sources

The greatest source on the history of H. Morris & Co., Ltd, belongs to the company archive. This was deposited by the late Neil B. Morris in the University of Glasgow, Business Records Archive around April 1995 in the move to Castlemilk, with an additional collection at Rutherglen. The records date from the early part of the twentieth century to 1975 and beyond. The records were moved from Milton Street by the Glasgow University archivists to the west end of Glasgow, before the old factory was demolished. Every nook and cranny of the old factory was searched by the archivists and this brought out something of value and that included documents on the Clydebank built "Empress of Britain's" furnishings stored in a false ceiling. The most valuable of the historical documents are the product line books drawn by Jack Cantle. These cover the early part of the twentieth century and are very beautiful, consisting of hand drawings and coloured images of the complete Morris range. There are two sides to a piece of paper and in this case it proved to be so – interworks memos had a message on one side, "...get McKay to send those papers out!" and the complete details of the production of the Armstrong Whitworth Albemarle aircraft on the other.

The William Weir archive was also a rich source of material as were the surviving papers from members of the Committee of Imperial Defence in the University of Glasgow Archive. The former Clyde shipbuilding and shipping company records were all consulted to correlate the ship furnishing work as was Lloyds List from 1900 to 1970. The list confirmed the shipbuilders building number for each vessel the company furnished. J. & G. Weir assisted with information on Cierva and the Weir helicopters as did Westland Helicopters. Many published sources were consulted at the Mitchell Library Glasgow. Periodicals included the complete run of the, "Cabinetmaker and Complete House Furnisher" from 1885,

Parliamentary and Ministerial documents as found in the published indexes and Hansard for both houses. Other periodicals included "Aeroplane," "Flight," "Aeronautics," "Shipbuilder," "Ships Monthly," "Shipbuilding and Shipping Record," transactions of the marine engineering societies, aeronautical societies and shipbuilding societies (including American societies) and all the local Glasgow newspapers. A complete set of intelligence reports on the German aircraft industry was made available from the Mitchell Library. This covered development of the German helicopter, jet engines and aeroplane plywood (These are the C.I.O.S./B.I.O.S./F.I.A.T. reports promulgated by the Combined Chiefs of Staff in Washington.) The papers of the Aeronautical Research Committee, Air Ministry, Ministry of Aircraft Production and Ministry of Supply were also consulted in relation to propeller manufacture, jet engine development, the use of aircraft materials, the de Havilland Mosquito, the Vickers Trans Sonic Missile and the tests on the Saunders Roe SR.A/1 jet fighter flying boat. The library also held information on the 1938 Empire Exhibition and the 1951 Festival of Britain and North American Supply. Also made available was the report of the Board of Trade into the industries of Imperial Russia. An in-depth European, United Kingdom and United States patent search was also manually carried out confirming dates and individual interests of H. Morris & Co., Ltd. Information was received from the Air Historical Branch, Ministry of Defence and the Royal Air Force Museum; I also received help from the Mosquito Museum at Salisbury Hall and the restoration team of a Bristol Beaufighter which was being restored to flight at Duxford. The former Ministry of Defence Pattern Room provided information on the production of the No III and No IV Lee Enfield Rifles and its projected replacement, helping to tabulate Morris production records and figures. I was also ably assisted by the Royal Aeronautical Society Librarian. It was the policy of the Ministry of Supply/Ministry of Production that all wartime production records be removed from the manufacturing companies who were involved in munitions work at the end of the 1939 – 45 war, Morris were very lucky to have some records of the company's wartime work from the archive.

The other great source of information was the oral tradition of the company and the memoirs of employees as well as those of the Morris family, which will be found in the first "Shavings for Breakfast."

The conclusions drawn from the statements, published facts and memoirs used in this publication are not necessary those of H. Morris & Co., Ltd., or its divisions in the Morris Furniture Group but are those of the persons who were interviewed or contacted. The texts on published sources and the conclusions given and used are those of file administrators or authors in official documents or authors in publications of those times. I have been given fullest access to surviving company papers, documents, files etc. I alone am responsible for the views, conclusions and statements expressed therein. Some data has been left intentionally blank. The author takes no responsibility for the conclusions drawn by the reader.

C. E. Mac Kay Saint Nazaire August 2007

Morris Furniture

Very little is known about the early life of the company founder Harris Morris, but what we do know is that he was born in the early eighteen eighties and the family came from the Russian European province of Saint Petersburg, which lies beside the three Baltic States of Estonia, Lithuania and Latvia. Dominating the province was the huge city of Saint Petersburg, with the other Russian province of Finland to the north. This was the Russia of the Tsars and the Tsar at this time was Tsar Alexander. Russia had not yet been industrialised on the scale of other European nations such as Britain and France. Russian industry was cottage based. But in the province of Petersburg there was one major national industry and that was the manufacture of plywood.

Plywood was exported all over Europe and a third of Britain's imports of plywood came from Saint Petersburg. When exported, plywood was packed and measured to "Saint Petersburg Standards." All European plywood was measured to this standard and it remained so right into the Twentieth Century. Plywood furniture had been experimented with and there was a working knowledge of its use in furniture, but at the Berlin Exhibition in 1885, the Russians exhibited a wardrobe made of plywood, which created a sensation. It was light strong and durable and economical to produce. It was also attractive to look at. But due to the size of the furniture industry in Russia and its inability to mass-produce no more was heard of the plywood wardrobe. There is an abundance of timber in Russia (29000 square miles of forest in 1881) and most of the working population were in wood related trades; carpenters, joiners, coopers, coachbuilders, all accustomed to working with the axe. In 1881 there were 154 cabinet making and joinery works in Russia with 4424 workers. Small firms in towns and villages all over Russia produced most of the furniture; doors, house frames, packing cases, barrels and mouldings. In the old

Russian Empire, the Department of Wood and Forests, through the Ministry of Imperial Domains, administered timber. The market for Russian wooden goods, including furniture, was not in Western Europe, but Persia, China and Bokhara, or sold at fairs all over Russia. A joiner could earn 45 roubles a month and a carpenter 41. Labourers or serfs could earn 176 roubles a year, but they could earn considerably more by carving wooden goods such as spoons, which could also be exported. Russian style furniture was heavy and wood stained; it consisted of footstools, chairs, sofas and tables. But by 1893 the market had shown a dramatic change with the export of bentwood chairs and furniture to Britain. The main port of entry for Russian furniture was London and the market was moving from the east to the west. Harris Morris was brought up in London. His father was described as a, "Traveller in soft goods." The family had arrived in London during one of the great upheavals in Russian history, which was towards the end of the reign of Alexander and the beginning of the reign of Nicholas II.

In his early teens Harry was apprenticed as a cabinetmaker to Harris Lebus, the biggest furniture and cabinetmaker in London. When he was twenty-one he completed his apprenticeship and left Lebus as a journeyman. From London he went to Newcastle and started working there as a cabinetmaker. Around 1906 Harry started up his business with three partners, J.N. Collins and glass merchants, F.H. Thompson and S. Bamford, both from Whitley Bay. They set up offices in Elswick Court, Newcastle, which was the business address of Collins. There are few records to indicate the type of work Harris Morris was doing, but glass was an integral part of cabinet – work and he also needed premises as a business address. The partnership seems to have worked well, but by January 1914 came a change in direction for Harry Morris and his partners.

Glasgow in 1914 was the second city of the British Empire. It was the hub of industry and the centre of commerce. What made Glasgow was the River Clyde with its heavy local industry. Steel and iron foundries fed the Clyde Shipyards and coal, which fed the iron works, was in abundance. The city had swallowed up minor, local towns such as Partick and Maryhill and its population could now be measured in the high hundreds of thousands. Glasgow boasted more than four hundred and ninety shipping companies including Cunard and the infamous White Star Line of "Titanic" fame. The shipping that arrived at Glasgow in 1914 came to a staggering 6.9 million tons. For the passenger ships of Cunard, the Anchor Line, The Alan Line, could be added cargo steamers, tramps, cross channel packets, coasters, lighters, barges and river steamers, which could be berthed at quays and docks. Glasgow could provide nearly eleven miles of accommodation for vessels.

On January 1st, 1914, Frank Thompson sold Harry Morris his premises at 147 Milton Street in Cowcaddens, Glasgow. The premises were part of the frontage of R. Laidlaw & Co, engineers and iron moulders and the sale cost Harry £2250. There he was to set up his business as a wholesale cabinetmaker in four small rooms. It is recorded in the Post Office Directory for Glasgow, 1914 – 15, that Harry Morris was one of 194 cabinetmakers in the city. On 2nd of March 1914, the partners met for the last time at Elswick Court, Newcastle, but this time there was a difference. Harry was proving that he was the driving force of the partnership when he called a meeting of the partners. The partners decided to form a company with each member having 100 shares. The company was to be called, "H. Morris & Co," the "& Co," were the three men from Newcastle, Bamford, Collins and Thompson. The latter became the chairman and the company secretary was Bamford. We now see the start of the Glasgow connection, when Thomas Brodie, of Saint Vincent Street, was appointed auditor. All the directors, except Harry Morris, were to serve a term of five years. Part of the agreement to set up the company was that Thompson and Bamford were to be the only suppliers of glass. One of the most interesting moves that Harry took was to have a telegram address, which was “Sheraton Glasgow.” At a stroke he was showing the cabinet industry where his company ethos was coming from.

Thomas Sheraton was born in the late Eighteenth Century in Stockton - on - Tees. Though he was trained as a cabinetmaker, he published, “The Cabinet-Maker and Upholsterers´ Drawing-Book,” originally issued in four parts between 1791 and 1794. His aim was teach the cabinetmaker how to build the most up-to-date designs. No piece of furniture can be found that can be ascribed to Sheraton. He did not invent the style of furniture named after him, though he certainly played a leading role in formulating it. The main centre for cabinet making in Scotland was at Lochwinnoch. There the Balfour family supplied most of Scotland with furnishings and fixtures for the post Victorian market. At that time there was a drive made for miniature cabinet work, centring on Jacobean pieces. Oak was the main wood used at that time, but oak had been used for ships in Britain for centuries and the wood was in short supply. The trade had to concentrate on using short pieces of oak which, when aged, resulted in splits along the length of the piece or in the backing panel. Due to lack of light in the homes of those times, furniture was characterised by its size and its dark colour. It was for this market that Harry Morris was to set up business and bring a profound change to the domestic and industrial market with his pleasing designs of a balanced nature using exotic woods from other lands.

William Beardmore had opened up a shipyard at Dalmuir near Clydebank. This move was primarily to gain work for his ironworks at Parkhead and to countermove any expansion of the Clydebank shipbuilding firm of John Brown. William Beardmore was in partnership with Vickers and Harry Morris had provided cabinetwork for Vickers in Tyneside. The Allan Line had ordered a liner from Dalmuir and she was called “Alsatian.” This vessel was to be used on the immigrant route from Glasgow to North America. H. Morris & Co. provided much of the cabin furnishings and panel work for her interior. But the days of peace were coming to an end and, by August 1914, the Royal Navy had requisitioned S.S. Alsatian as H.M.S. Alsatian. She was to be the flagship of the blockade of Germany on the contraband runs for much of the Great War.

There is little evidence in the company records to show any major war work in the first few years of the war. When the Ministry of Munitions was formed in March 1917 it was approached by the company for war work, but none was forthcoming. Later, as the war progressed, the company received orders for ammunition boxes for heavy artillery and orders to recondition returned ammunition boxes from the field. We do know that there was extensive work in the shipyards to provide panelling and cabinetwork on most of the Clyde built warships. There was also the new infant aircraft industry, which was growing on Clydeside. Many aircraft, such as the Airco D. H. 9, required plywood in their construction – a ready market for H. Morris & Co. From the company minutes we find that the large government contracts were secure by 1918 and that Harry Morris was in total control of the situation.

In October 1917, Harris Morris took the Oath of Allegiance and became a British subject. Events in Russia, culminating in the Revolution, had changed the international scene and the new Soviet Union was not recognised by the Allied Powers.

An extraordinary general meeting of shareholders was called at Newcastle in March 1918. At the meeting Harry introduced his wife, Dolly, to the other shareholders. Jack Collins had died in 1916 and his two brothers, Abraham and Isaac Collins, took up his shareholding. It was agreed that Dolly should receive 300 shares and that Bamford, Thompson and the Collins brothers would retire, leaving Harry Morris in control of the company. The brothers transferred their shareholding to Harry and left the company. Harry Morris formed the

Milton Property Company and this company became the owner of the Milton Street works. R. Laidlaw moved away and the company took over the remaining building, moving in from the small Milton Street frontage in 1920. The Board of Trade defined cabinetmakers as a business producing cabinet goods: wardrobes, dressing tables, tallboys, bedside cabinets, sideboards, kitchen cabinets, bureaux, bookcases and bureaux– bookcases. Clearly H. Morris & Co met this definition, but it could also be added they manufactured panelling to a high standard. The first major post – war contract was to provide cabinets for London County Council. This was a follow – on contract for the first major, post – war, contract, and the refurbishment of S.S. Alsatian. The Allan line had been bought out by Canadian Pacific, and the Alsatian was renamed, "Empress of France." When the vessel was refurbished, all her wartime fixtures and fittings were removed by Beardmore for the North Atlantic run to Montreal. The Morris Company were prime contractors for the refurbishment and the first link with Canadian Pacific was made with the contract on this vessel. After this came the contract to furnish the Gleneagles Hotel in 1923. Gleneagles is a luxury hotel in Auchterarder, in Perth and Kinross, Scotland. It was opened in 1924 by the former Caledonian Railway and it was known as, "The Riviera of the Highlands". The hotel was designed in the style of a French chateau and Capability Brown, the celebrated 18th century landscape gardener, inspired the landscaping, which is set in 850 acres of Scottish scenery. When it was built in 1923-24 the prime contractor for furnishings and fittings was H. Morris & Co. (Today it has 213 rooms and 15 suites)

It was in 1922 that the company set up its drawing office. From Bath Cabinetmakers came Jack Cantle who stayed with the company for more than 50 years. It was in that year that Harry asked the company secretary to define his contract and Dolly Hetty Morris became his business partner. This was ratified at a meeting in the family home, Bentinck Street (named after the 1835 MP) where the only shareholder present was Harry Morris. Harry also wanted to branch out as a wholesale cabinetmaker, but the company's Articles of Association had forbidden this. Harry Morris asked the company secretary to attend to this matter, which he did. H. Morris & Co became a wholesale cabinet making company in Glasgow.

Following "Alsatian," London County Council and Gleneagles came two unusual contracts. First was the upholstery work on the Beardmore taxi, which were being manufactured at Anniesland and secondly work on the Bennie Railplane. This was a rapid passenger elevated railway system proposed by George Bennie, to the London and North Eastern Railway Company. An aluminium car (bullet shaped with a propeller at each end) ran on an elevated trackway at Glasgow and Milngavie Junction Railway. It was constructed at Dalmuir by Beardmore and the doors, coachwork and internal fittings, were fitted by Morris. The original single chair design followed closely the tub chair design so familiar in ships. Sadly for George Bennie the idea was not taken up and the whole structure lay forlorn for many years and was demolished around 1956.

During 1923 there was a parliamentary investigation into the furniture industry. It was proving hard for the consumer to actually trace the manufacturer of household furniture. The investigation found that, due to sub contract work, it was very difficult, if not impossible to trace the manufacturer from the outlet shop. They also found that only three manufacturers had a trademark and one of them was H. Morris & Co. For a trademark the company had picked an unusual logo, a figure actually planing a piece of wood. The image was based on George Patterson, a long time and loyal employee of the company. The mark was issued on 11th October 1923. The logo was attached to a pin and pressed into the cabinetwork such as a corner of a wardrobe or the inside of the drawer in a chest.

By 1925 the shape of the company was starting to crystallise. There was a drawing office, which prepared product lines for hotels, ships and miscellaneous contracts, blacksmiths, for attending to the metalwork in the new cinema chair market, a garage for trucks, storage areas for furniture wood and a plywood manufacturing base. Harry Morris controlled all of this. He knew every employee (he in fact had hired them himself), and he knew every piece of machinery. He knew every contractor and when anything went wrong he was the first man called in. He inspected every piece of furniture and if it were not up to standard it would not go out of the factory. By 1925 he was employing about 175 employees in Milton Street.

During 1928 the company purchased the remaining property in Milton Street. The economic collapse of the stock market and the decline in the economy in 1929 did not affect H. Morris. This was probably due to the nature of his business. He was supplying to the hotels, shipyards and completing wholesale cabinetwork for the suppliers. This was a very difficult operation to fulfil for, in reality, the company was the Glasgow "New Boy" and he was in competition with such established firms as Wylie and Lochhead, Rowan and Boden and the multitude of cabinetmakers that grew on every Glasgow street corner. What was known about H. Morris & Co, was its ability to give the customer quality goods which would last and were easy to maintain. But there was a period of depression in the shipyards, especially with John Brown at Clydebank. In the early twenties, the only major vessel being built at Clydebank was the Cunarder, S.S. Franconia, Morris supplying panelling and cabin furniture for this vessel. There is no cabin work recorded for the next two major Clydebank vessels St Julian and St Hellier of 1926. By 1928 major ship work was starting to come in for Milton Street. Renewing its ties with Beardmore, since the completion of work on three vessels of the Lloyd Sabuado Line in the early twenties, contract work was received for the Duchess of Athol. This included panelling for the public areas, with furniture and cabin work. This latter work included wardrobes, beds, and occasional tables, panelling and standing lamps. The Duchess of Athol was built for Canadian Pacific to exacting standards. Canadian Pacific took responsibility in all their passenger vessels for furnishing. This meant that the builder made the vessel to the bare hull and it was up to Canadian Pacific to furnish the vessel on completion.

Canadian Pacific contacted H. Morris & Co in early 1927. A false cabin was built on the factory floor and product books readied. Due to the nature of the work personnel in the drawing office had to be aware of calculating the stress on the hull from the twisting and turning of the hull in the ocean and more importantly stress in the cabins. After the cabin was inspected the product books would be consulted. Panelling was chosen, as were all the fixtures and fittings for the cabins. Then Canadian Pacific changed its publicity for its ships. There were no longer cabins but apartments; apartments, finished to the highest of standards. There were three other Duchess vessels ordered for the Clyde but Beardmore did not attract these orders due to their financial predicament. The orders went to their arch rivals John Brown. The three vessels were the Duchess of Bedford, Duchess of York and the Duchess of Richmond being built between 1928 and 1929. At the time they were the largest cabin vessels ever built for the sea voyage to Montreal. With a length of 600ft they accommodated about 1500 passengers in cabin class, tourist class and third class and they weighed about 20000 tons. Again these vessels were built to exacting standards with the public areas and the cabins being furnished and panelled by Morris.

While the Duchess vessels were being built an order to furnish three further vessels for cabin work and furnishing was received at Milton Street. The first vessel was The RMS (Royal Mail Ship) Rangitane; she was built as yard number 522 and launched on Monday 27th May 1929. She was one of three sister 'Rangi' ships built for the London registered New Zealand Shipping Company (NZSC), the others being the Rangitiki and Rangitata. Individually they were nicknamed “Tane,” “Tiki” and “Tata.” Each was 16,700 tons and could carry nearly 600 passengers, 200 crew and substantial cargo and all were built specifically for the England-New Zealand run. They had twin propellers powered by Brown Sulzer diesel engines with a total output of 9,300HP. Rangitane became a war loss, sunk by a raider in 1940, Rangitiki and the Rangitata served as troop transports from 1941 to 1945. Another shipbuilder, Fairfield Shipbuilding & Engineering, received an order to build the Empress of Japan in June 1928. She was built to run the service between Vancouver and Yokohama. This ship was a departure for Morris because the Fairfield shipyard was on the south of the Clyde whereas the other yards Morris worked in were to the north of the Clyde. Morris supplied all her panelling and furnishings for the cabin class accommodation. There were 399 cabins fitted with bedsteads and two deluxe suites had a veranda sitting room. Most of her panelling was in ash and walnut. She was launched on 17 December 1929 and made her maiden voyage to Quebec in June 1930. Due to war conditions she changed her name to the Empress of Scotland in 1942. She sailed as a troopship from July 1940 to May 1948 with the same captain; she had been at war so long that many of her crew literally had holes in their shoes due to her turnaround at ports. But the personal kindness of an American general meant that the crew were promptly kitted out with new shoes.

Further Canadian Pacific vessels were ordered from John Brown and one such was yard number 530, “The Empress of Britain.” This ship was to be the most luxurious liner ever built. The Canadian Pacific Company designed her in conjunction with J.M. McNeil of John Brown; even Mr Johnson of Canadian Pacific was left to design her boilers. She was laid down on 28th November 1928 and was the last word in shipbuilding in those times. P.A. Staynes and A.H. Jones were the decorative artists for Canadian Pacific, but for such a high profile vessel as this one, five artists were commissioned to do the work. Staynes and Jones were left to design the corridors and public areas, Sir John Lavery designed the Empress Room, Heath Robinson designed the Knickerbocker Bar and Sir Charles Allom designed the large Mayfair Lounge. The Cathay Lounge was designed by Edmund Dulac and Frank Brangwyn the Salle Jacques Cartier. The Mayfair Lounge was decorated with rich walnut panelling inlaid with silver. Around the lounge were writing tables, sofas, card tables and chairs. Dulac designed the Cathay Lounge with panelling of grey ash and the floor was of Macassar ebony and patterned oak. (Macassar ebony was a favourite of Harry Morris.) Ebony furniture was dotted around the lounge. The Salle Jacques Cartier, with walls of light oak, accommodated 454 diners, while two smaller rooms, Wolf and Montcalm, accommodated 36. Mahogany and oak panelling was installed throughout the ship, with some areas having soundproofing. HRH Prince of Wales launched her on 11th June 1930 and, unusually for a ship in those times, most of her internal furnishings were complete when she entered the Clyde. When she reached Montreal on her maiden voyage she had taken 5 days 13 hours and 25 minutes to reach Wolf's Cove from Southampton. She was hailed as the wonder of the age and the quality of her accommodation became the byword for Atlantic Liners. Morris was involved in all aspects of her furnishings, supplying everything from mirrors to double beds and no matter what Canadian Pacific said of their apartments, to the passengers they were "Cabins." The Empress of Britain set the fashion for cruise ships with the use of large-scale veneered panels, manufactured by Morris and this was to be continued in further luxury vessels.

The Empress of Britain could be converted to a cruise ship by removal of two of her propellers at dry-dock. During June 1939 she became a royal ship when she embarked King George VI and Queen Elizabeth returning from America in a very happy voyage. Except for two new beds in the King and Queen's apartments regular Morris furniture was used in all the other cabins. When war broke out she was requisitioned as a troopship. On 26 October 1940 she was attacked by aircraft, the Mayfair Lounge was bombed and she quickly became ablaze. A few days later her glowing hulk was torpedoed and she was sunk, the largest war loss of a civilian vessel at sea. Canadian Pacific and the whole of America mourned her loss; the "New York Times" said of her, "No ship ever fitted her name more than the Empress of Britain. She was, indeed, an empress, with pride and grace and dignity in every inch of her." The King and Queen sent their condolences to Canadian Pacific, expressing their sympathy at the loss of such a fine ship.

With the completion of the Empress of Britain and the prospect of high volume work from the shipyards and the expanding hotel work, Harry Morris extended the factory. In 1932, a new factory was built beside the old one and the two were combined to cover the vast corner site of Milton Street. This was an exciting move, but the merging of the new building with the old building meant that with the combination of the two on the Milton Street site, the combined factory had reached the limits of expansion. No further development was possible at Milton Street and that situation was to last until the nineteen eighties. Harry Morris was a familiar figure around the factory and the shipyards of Clydeside with his flat cap and dustcoat and he was now going to enter the busiest time for his business. John Brown at Clydebank had received an order from the Cunard Line for a vessel listed as yard number 534. This ship was to run across the Atlantic between New York and Southampton. Brown's introduced Harry Morris to B.P. Camp who was the furnishing superintendent of the Cunard Line and who had been involved in the outfitting of Aquitania and other ships. 534 was to be the biggest ship in the world, weighing in at 81237 tons with an overall length of 1018 feet. There was to be a garage, a telephone exchange with 700 lines with several miles of telephone wire, 21 electric lifts and even a small chapel and altar. She was to carry 2000 passengers in the familiar cabin style of the Empress of Britain, cabin, tourist and third class. Unusually for a ship in those days, 534 was to be all-electric. Camp arranged with Harry Morris the terms of the work on 534. H. Morris & Co were given the responsibility for most of the panelling on the ship, the furnishings and fittings for the cabin class accommodation or staterooms and any intricate and detailed work that Cunard seemed fit to authorise. To quote Harry Morris, “The work we have done for this liner was the largest amount of woodwork done by any contractor in the woodworking industry and embraced all the cabin class accommodation." The cabins were decorated with over fifty varieties of wood measuring over 22 acres. The walls were of ivory-white sycamore with a faint ripple of white. Other cabins could have maple, African cherry, pearwood, Pacific myrtle or English yew. The furniture was of blended woods, tall mirrors and long-glass wardrobes, which lit when you opened the door. The writing tables were beautifully decorated and softly lit and there were wide beds, which became a divan by day. You could have a private dining room as well as a sitting room and servant's quarters could be provided. 534 was lavish on a scale never seen before and tastefully decorated. To quote Cunard, "No expense having been spared to make this ship the most beautiful and comfortable afloat."

Due to the depression work on 534 stopped and she lay a rusting hulk at Clydebank. The government stepped in and after an enquiry by Lord Weir, Cunard merged with the White Star Line. Other firms were feeling the effects of the Depression, but Harry Morris was able to report to his bankers that he was very busy and that his accounts were improving. The merger of the two companies and the influx of government money meant that work on 534 could resume. On the 30 September 1934, Her Majesty, Queen Mary, in the presence of King George V and Edward, Prince of Wales, launched 534, naming her "Queen Mary." It took eighteen months to fit her out with 7000 men working on her to get her ready for her maiden voyage, ready to meet the spring tide. On 24 March 1936, Queen Mary left Clydebank fitting - out basin grounding for an hour on a sandbank, due to her enormous size. She sailed into history on 27 May 1936 when she left Southampton for New York. For years she proved to Cunard engineers that she would be a difficult ship and tiring to run.

Even with the depression, British shipping was still carrying 29% of the world's trade. Measures had been taken to close unprofitable shipyards and make the survivors profitable. Dalmuir was closed down, as were other Clyde yards and it was left to John Brown, Fairfield, Stephens's of Linthouse, Barclay Curle etc. to continue the tradition of Clyde shipbuilding. Vickers still had their two yards at Barrow and Newcastle and had attracted two orders for the Furness - Withy Line, the two ships being the Queen of Bermuda and the Monarch of Bermuda. The 22,500-ton Queen of Bermuda was one of the great liners of the 1930s. She was completed in 1933 at the Vickers-Armstrong Yard at Barrow-in-Furness and, together with her near sister, Monarch of Bermuda of 1931, added great luxury to the Bermuda cruise trade. 550 feet long, with splendid public rooms, a large main restaurant, an indoor pool and spacious sports and sunning decks, they boasted a great novelty for that era, a private bathroom for every cabin. All the cabin work was completed by Morris, as was the panelling throughout both vessels. The Queen of Bermuda became known as the biggest floating ballroom in the world and sailed on the Caribbean route between New York and Bermuda. In an interesting little twist of history, while Furness-Withy Line awaited delivery of the Queen of Bermuda in 1933, they chartered the Duchess of Bedford from Canadian Pacific Lines, another Morris ship. They were called the "Honeymooners ships," a voyage costing as little as $62 for a six day round trip. Both vessels were requisitioned in 1939 as troopships and were involved in every major Allied landing surviving the war to sail into peacetime.

The Anchor Line had placed orders with the Fairfield Company of Govan; the ships were Circassia and Cilicia. The work to be completed on these two vessels was small in comparison with the other major contracts of those times. Cabin work was completed for the radio-officer and the captain. Little can be established in the way of panelling or ship furnishing for these ships, so it may be supposed that this work was carried out as a sub-contract for one of the other ship outfitters on the Clyde such as Wylie & Lochhead or Rowan and Boden. Fitting out part of the Cameronia was another contract that is difficult to trace, but certainly Morris did work on this vessel. Cilicia made her maiden voyage to Karachi and Bombay on May 14th 1938. In 1939 she was requisitioned as an armed merchant cruiser. At the end the war she was released back to Anchor Line service for her peacetime role, having transported 16,035 troops and prisoners of war. Cameronia was used to transport Polish gold to New York in 1940, the bullion being sealed by the button of the Anchor Line set in red wax. The button came from the captain's tunic. Circassia survived the war to be the first ship to sail to Saigon in French Vietnam carrying Japanese prisoners of war for police duties. Her captain, David Bone, was the brother of Muirhead Bone, celebrated ship artist. These vessels were very popular with their crews and passengers alike, due to the standard of accommodation and their sea keeping qualities.

During 1938 Neil Balfour Morris joined the company. Born in the west end of Glasgow he was educated at Hillhead High School. When he left school he was originally apprenticed to a mechanical engineer, but joined his father at Milton Street. He was selected to travel around the continent inspecting the furniture industry of Holland and Belgium and ultimately went to work in plywood mills in Germany. He had witnessed Hanna Reitsch flying the Focke Achgelis helicopter inside the dome of the Deutschland Halle and was enthusiastic about vertical flight. From Germany he went to Le Havre, then to Paris, to study decorative veneers. From France he went to Switzerland, where he learned Swiss plywood techniques.

Neil Morris was very interested in the New York World's Fair. Earlier, in 1938, Glasgow had hosted the Empire Exhibition with Morris once more working flat out to decorate many of the colonial and dominion exhibits. Many were the visitors from New York and many were the invitations to visit New York for the 1939 World's Fair. Neil Morris received many invitations to go to New York and this attracted the attention of the Air Ministry. Would Neil B. Morris have a look at the woodworking capacity of the United States, particularly the manufacture of wood glue for the ministry? Yes he would.

Sailing by Canadian Pacific to Montreal, Neil Morris went to New York in June 1939. He went all around the eastern United States, particularly Grand Rapids looking at the woodworking techniques of the American woodworking industry. From the east coast he headed by aeroplane to Washington State to look at the American forestry programme. He returned to New York and then sailed back to Britain on the eve of the Second World War. With him he brought all the reports he could find on American wood glues, especially those produced by American Forest Products. In his report to the Air Ministry he found that the American woodworking industry was not as efficient as in Britain. It did have a production capacity for specialised work, but was far too fragmented. However the advances they had made in wood glues and bonding were worthy of investigation.

By 1939 it was clear that war was inevitable and the Clyde Shipyards were now working on naval contracts for the Admiralty. Orders had been received for battleships, aircraft carriers, cruisers, destroyers and frigates and Harry Morris wanted to get into this type of naval work. In 1936 he had written to the Admiralty specifying the type of work that Morris Furniture had carried out for the civilian shipbuilding industry. But there was one outstanding order in Clydebank that was building and that was the Queen Elizabeth. She was to be the ultimate Cunarder, with Morris being involved in all departments, from the doctor's surgery to the captain’s cabin, cabin class, tourist class, third class, panels, panelling, lifts, stairs, fireplaces and windows. Even the humble coat hanger had its place in the Morris inventory. But Queen Elizabeth was to see no hail of publicity on her maiden voyage. Dropping her cables, the grey Queen Elizabeth calmly set sail down the River Clyde and headed for New York. Britain was at war.

The Beardmore taxi was built in Anniesland in Glasgow's west end, Morris completed extensive coachwork in the interiors, seats, door panelling, and headlining for the roof.

The interior of the Bennie Railplane. Notice that the seats are similar to tub seats found in contemporary ships.

The experimental Bennie Railplane at Milngavie in the nineteen twenties. When completed at Beardmores in Dalmuir and taken to Milngavie, Morris installed the seats and interior work.

The front cover of Bennie's publicity booklet showing his vision for the future

The company logo

Entering service in 1923 Morris furnished the Franconia for the Cunard Line. She was available for cruising as well as the American liner service

The Beardmore built Conte Biancamo for an Italian Shipping company. Though Morris provided the cabin furnishing and panelling, many Italian craftsmen were brought in to do the detailed artwork such as cornices and ceiling work.

Another Italian liner Morris furnished was the Beardmore built ConteRosso. Many of these fine vessels became war losses in the Second World War.

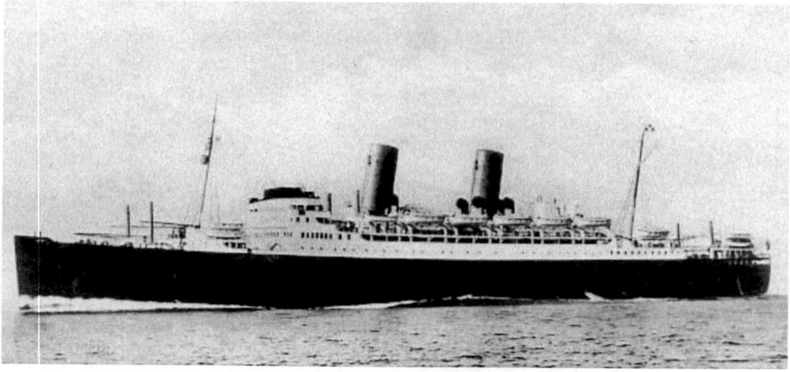

The Duchess of York established Morris links with Canadian Pacific. Morris panelled the public areas and supplied most of the cabin furniture

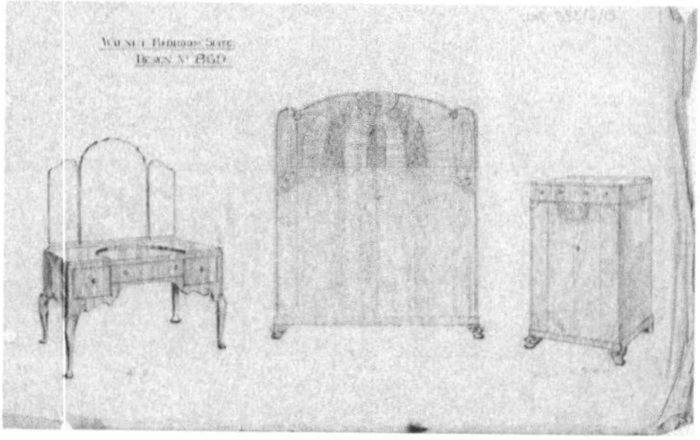

A dresser and wardrobe in Georgian style set in walnut and mahogany dated from 1923. The designer of all these pieces was Jack Cantle.

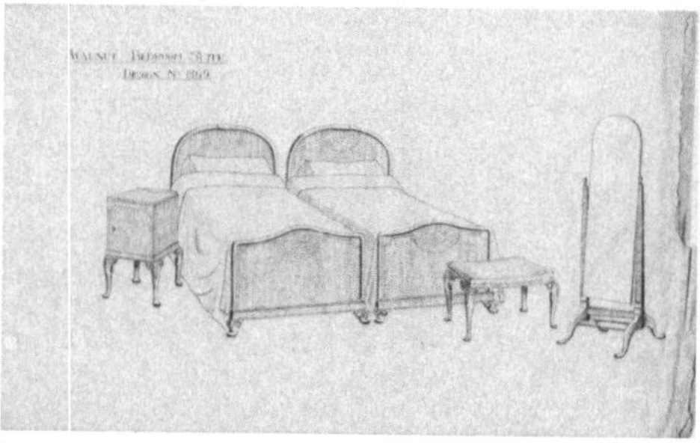

Single beds etc. for hotels and ships

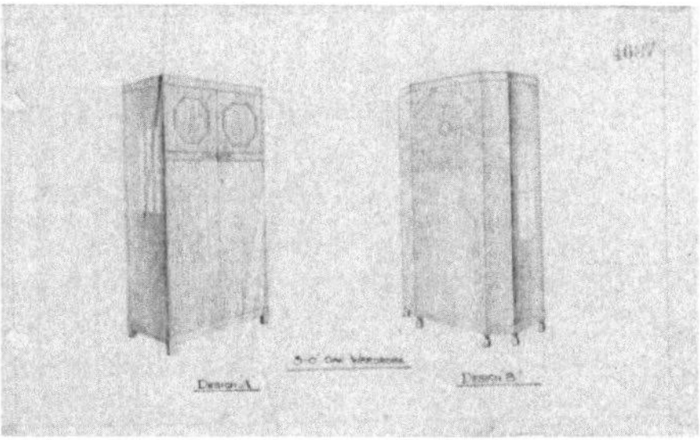

Product 4687, design "A" and design "B" which could be adapted for any use.

Product line 4616 from the product line book of the nineteen twenties.

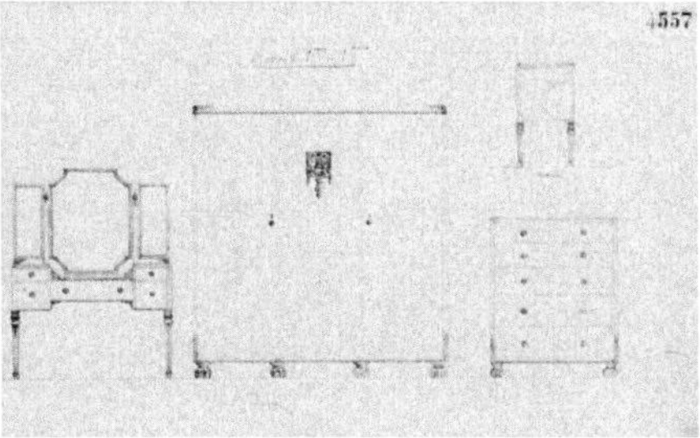

Set in oak, product 4557, dresser wardrobe, chest and pedestal

A modern dresser with stool and swivel mirror

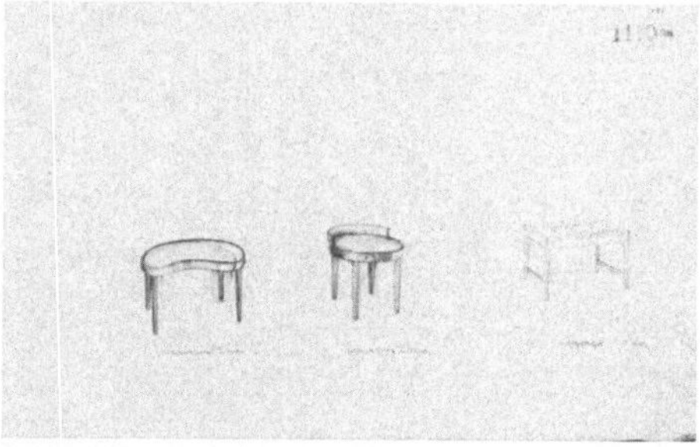

A variety of stools, one padded with Dunlopillo, for the hotel industry

The entrance Carcase for a lift.

A single bed showing a variety of polished wood ends.

Adverts for Canadian Pacific, including the 1938 world cruise of the Empress of Britain

The 1938 world cruise of the Empress of Britain. For 128 days passengers could enjoy the voyage of a lifetime

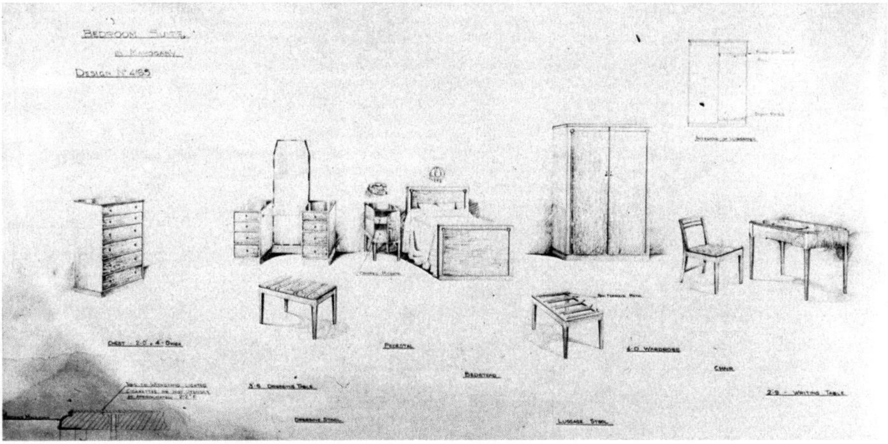

Hotel suite 4165 in mahogany. Five complete suites were finished per week, which could be extended to ten per week depending on the demand

The old North British Hotel off Glasgow's George Square, beside Queen Street Station,

The Hope Street entrance to the Central Hotel

The now demolished Saint Enoch Hotel, which was furnished by Morris. The hotel also served as naval headquarters during the Second World War

The Canadian Pacific liner Duchess of Richmond which Morris furnished for the builder, John Brown at Clydebank

Rangitiki built at Clydebank for the New Zealand Shipping Company in wartime colours. She was a very popular ship and Morris refitted her post-war at Clydebank.

The Empress of Scotland, formerly the Empress of Japan; here she is docked at a port in America for a voyage to the Clyde.

An impressive view of the Empress of Scotland. She was sold post-war to a German Company, but sadly was lost in a fire in the nineteen fifties.

Yard number for the Empress of Britain was 530. She set the standard in luxury for a liner, which has never been surpassed

CAPTAIN'S BEDROOM
CAPTAIN'S BRIDGE

The Captain's bedroom on board the Empress of Britain. All Morris products were used in its completion

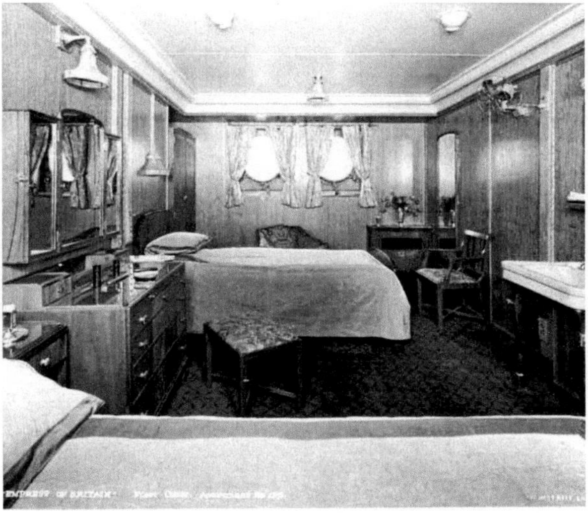

FIRST CLASS STATEROOM Nº 139

The First Class Stateroom set a standard of comfort for the North Atlantic runs. The Royal Family on their American visit used these staterooms in 1939

Special room no 17, again finished to a new standard in comfort

The Empress of Britain sailing on the calm of the Saint Lawrence, June 1931, her lines reflecting on the water, a classic for the age.

Advertising the "White Empresses"

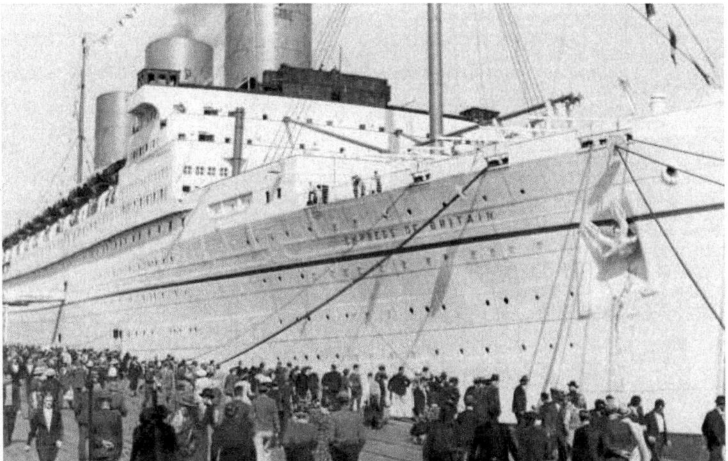

The "White Empress" alongside in Australia

The Empress of Britain on fire after being bombed by the Luftwaffe. Her bridge superstructure has collapsed and her hull is glowing from the internal fires

Exterior of the Milton Street factory, built in the nineteen thirties

The main door at Milton Street showing the three storey design as adopted by Harry Morris

Queen Mary at sea off Australia

The Queen Mary on a voyage to New York

The Queen Mary at New York. This was the biggest ship contract the company ever had for one vessel

The Queen Mary returns to New York in June 1945 with 16000, troops a very conservative estimate. The Queen Mary had sailed from Gourock on her first post – war voyage with American servicemen

Queen Mary taking a high sea at speed, while on a trooping voyage. She was so fast she never sailed in convoy. Her furniture had been off-loaded to South Africa, the United States and Australia, before she undertook her wartime trooping. In June 1943 she carried 23000 soldiers to Gourock, Scotland, the largest ever passenger haul

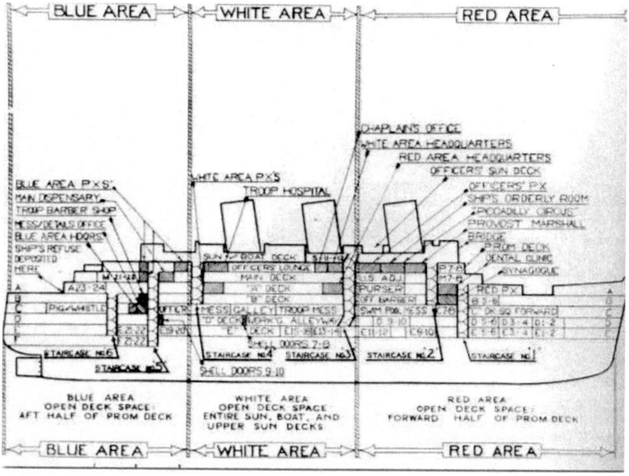

Queen Mary trooping boarding card for American servicemen. Her accommodation had been split into blue, white and red areas.

For the Anchor Line, Fairfield of Govan built the Circassia, Morris were involved in cabin work for the captain's cabin and the radio officer's room

The Monarch of Bermuda

Queen of Bermuda first class dining salon

Queen of Bermuda

The first Lord Weir, who investigated North Atlantic Shipping

Queen Elizabeth on the stocks at Clydebank. When she was completed she was the biggest ship in the world.

Queen Elizabeth on her maiden voyage to New York

1938 Empire Exhibition at Bellahouston Park with Tait's Tower to the right

A Wartime Interlude

Much changed with the outbreak of the Second World War at Milton Street and the world. During 1937 Harry Morris moved from Glasgow's west end to Titwood Road on the south side of the city. He drove a 1937 Wolseley and he was quick to order a set of tyres for the car just after the declaration of war. The factory was now working at capacity with over 300 employees, many with years of service with the company.

Work started on the King George V battleship, H.M.S. Duke of York (building at Clydebank), fitting the admiral's accommodation, including electric fireplace, wardrobes, tables and a large writing table and the officers' mess and quarters. A crack warship renowned for the accuracy of her gunnery, she sank the German battleship, Scharnhorst, in the autumn of 1943 and was the representative of the British Pacific Fleet at the surrender of Japan with King George V and the U.S.S. Missouri in 1945. At Tokyo Bay Admiral Sir Bruce Fraser offered Duke of York's furniture from King George V, to General MacArthur for the surrender ceremony on the U.S.S. Missouri. McArthur is said to have declined the offer due to the size of the table but used the chairs instead. The table used was an ordinary GS table. The furniture of the Duke of York is reputed to be in a United States Navy Museum. For H.M.S. Fiji (a crown colony cruiser) Morris furnished the captain's cabin, officers' mess and quarters and all her doors, which were specified as fire doors. A destroyer, which became the Polish Piorun and two Hunt class destroyers received fire doors and furnishings for all the crew accommodation. Lastly, Morris received an order for cabin work on the last British battleship, H.M.S. Vanguard, which was launched in 1944. Most of the wartime activity was taken up with work for the Ministry of Production, Ministry of Supply and Ministry of Aircraft Production, the Air Ministry (outfitting the officers' mess at RAF

Uxbridge - Fighter Command), the Admiralty and the War Office. At his own request, Harry Morris approached the LMS Hotel Group for an order to supply limited furniture for the group. His hope was to keep the furniture skills of the workforce at a peak. There were also orders received from Rolls Royce at Hillington, to supply packing cases for Rolls Royce Merlin aero engines and crankshafts. These orders existed throughout the war, as was the supply of ammunition boxes to carry .303 small arms ammunition and cordite trays for Ardeer. Another interesting order was to provide Frank Whittle with a wooden compressor for one of his intermediate gas turbine engines. The concept worked, but at such a high speed the laminated wood blades broke and would not have proved successful in service. (This idea of a composite jet engine was similar to the use of carbon fibre blades in the Rolls Royce RB211 engine of 1970-71, which also failed.)

But there was an artist whose needs were stronger than the war effort and that was Benno Schotz. Benno was an established sculptor whose expression was in bronze. Due to wartime restrictions he gave up his profession to work in John Browns at Clydebank. Benno had a desire to set up an exhibition in the Royal Scottish Academy with a new piece as a centre of his expression. Unusually, he departed from his love of bronze, to a figure carved in wood. His inspiration was Psalm 119, quoted to him by his friend Dr Honeyman. When he saw the wooden figure he exclaimed, "Ah, unto the hills I lift mine eyes," capturing perfectly Benno's mood and thought. Reflecting the carnage of the times the female figure was depicted with its head facing heavenwards, its arms clenched tightly against its body and the shoulders raised, with the ears turned out for deliverance. Benno wanted a piece of wood that had a special twist to it which he could form into the shape that stood out in his mind. Taking the tram, he headed for Milton Street and Harry Morris and the two of them scoured the old left - over wood from pre-war furniture and fittings that littered the back of the motor yard. To their surprise they found a four-foot piece of wood exactly what Benno was searching for, even to the desired shape and twist that he needed for this work. The wood was Lignum Vitae, or ironwood, three times as hard as oak and used in the shipbuilding industry to support bearings and rollers and also to make mallet heads. Lignum Vitae is a reddish brown colour when freshly cut with pale yellow sapwood. As it ages it becomes green with black patching. The grain is highly interlocked, making it difficult to work with edge tools, but it machines well and takes a high polish. It is a remarkably good wood for turning. When Benno finished the piece he invited Harry Morris to see the completed work and he brought with him Neil Morris to show that there was another man in Glasgow who loved wood. Neil asked Benno if he would fill in the imperfections of the wood. Benno said, "No!" He was leaving the imperfections (shakes) just where they were. Neil added, "Quite right, they are the scars of life." Benno realised that Neil Morris loved more than just wood.

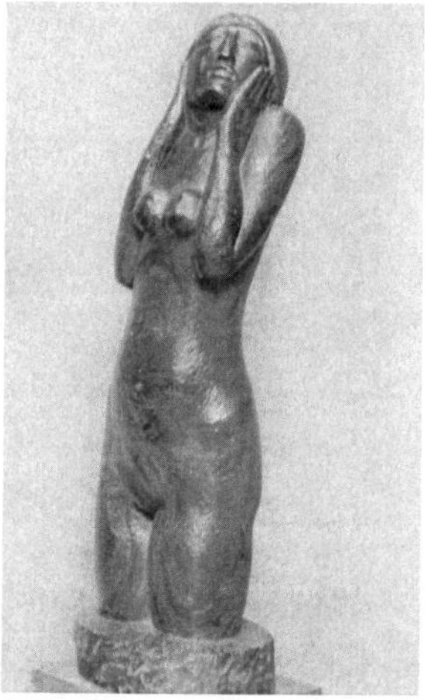

Lament by Benno Schotz in ironwood, taken from the Morris yard at Milton Street 1943-44.

Duke of York in Icelandic waters protecting the Arctic convoys and covering the attack on the German warship Tirpitz

Morris provided the fire doors as well as cabin work for the cruiser H.M.S. Fiji

Morris completed all the cabin work for the last British battleship, HMS Vanguard, seen from the carrier Implacable during "Operation Two Step," also seen here in post war views

Port view of Vanguard on trials on the Clyde

Vanguard on trials off the Ayrshire coast

The end of Vanguard, Faslane 1960

Morris built large quantities of packing cases for Rolls – Royce at Hillington. Many of the cases were for Merlin crankshafts. Here is a Rolls – Royce Griffon

Railway Hotels

"The L.M.S. Hotel Services and, subsequently, British Transport Hotels, had a long association with Morris of Glasgow. I was lucky to be part of that association for over thirty years, first as a junior housekeeper at the Central Hotel, Glasgow, then as an assistant in the Domestic Services Department of the L.M.S. Hotel Services and, as Decorating and Furnishing Manager in the Works Department of British Transport Hotels. Although I only met Mr. Morris (Neil's Father) once or twice, I didn't really know him ... only the results of some of his labours. I have a clear recollection of him coming into the Central Hotel in 1937 to see Arthur Towle, controller of the L.M.S. Hotels. Any visit by the controller to a hotel was not easily forgotten ... even the tops of the high lavatory cisterns were specially dusted. He was a very tall man and set high standards. I looked from afar at this formidable figure strolling along the wide corridor with a short stout man taking two steps to his one, but keeping up. Although I wasn't there at the actual meeting I know from later experience that the controller, for his comfort, would loll full length in one of the vast easy chairs, which were part of the standard equipment of all L.M.S. sitting rooms. Mr Morris would have to perch uncomfortably on the edge of his chair ... if he'd copied Mr Towle he would have been less comfortable. Either way Mr. Towle would have the advantage. The fact that he drove a hard bargain on everything he dealt with may have been helped by these tactics. The two gentlemen had dealings over a long period and must have had many meetings, since Morris of Glasgow provided some of the suites for the Gleneagles Hotel, which opened in 1924 and for several other hotels in the group. I cannot confirm how many of the fine dark mahogany pieces in the main bedrooms at Gleneagles were made by Morris, but we were always under the impression that the furniture, so well suited to guests who brought vast quantities of luggage and which was still in the best rooms when the hotel was sold in 1982, was made by Morris. The furniture ordered in 1937 was very different. The Empire Exhibition was taking place in Glasgow in 1938 and bedrooms had to be updated. The requirements for bedroom furniture had changed considerably since 1923 when the hotel was built and the only new furniture had been provided in the 1920s for suite bedrooms. Details were carefully worked out and drawings produced, and, eventually, delivery effected in the spring of 1938. Mr. Towle drove a hard bargain on everything he dealt with, shelves, ladies section, long enough to take long evening frocks, tie rack, all things taken for granted today ... but very different from the 1923 type, with mirror door, long drawer at the bottom, hat cupboard, and drawers at the side. Dressing-tables had long mirrors, to enable even Arthur Towle to see himself full-length; two sets of drawers the width of a man's shirt, providing a knee-hole for ladies to sit at. These replaced dressing chests with five drawers, swing mirror, two jewel drawers and a white linen runner, changed daily, not to mention clean white paper lining each drawer. Low bed-ends replaced high brass beds, twins replaced double beds (all hotel beds were originally double and often placed against the wall so that one had to climb over the other person to get in and out of bed.) The new beds, of course, were 6' 3" or 4" long. A neat glass-topped cupboard replaced the old chamber cupboards beside the beds, with a wardrobe with panelled doors, piano hinges, bronze handles and press button openings (no keys ...they get lost), shelves instead of drawers and men's hanging space below the space for telephone books. The problem of the telephone v. the morning tea-tray was, however, not improved ... an old-fashioned candlestick phone is easier to grab and put on the floor while balancing a tray!

Simple luggage stools, small writing tables with one drawer, chairs and dressing stools with drop-in seats (minimum covering material) and excellent space-saving tallboys, completed the set. These latter, due to the war and people's changing habits, were the only items needed. With all these requirements Messrs Morris produced a co-ordinated suite of bedroom furniture in Honduras mahogany that transformed the rooms, pleased the guests and delighted the staff. Much of the furniture is probably in use still somewhere today, having had practically no maintenance for nearly fifty years. Of course 1 did not realise all this at the time that I was helping to dump massive immovable wardrobes and dressing chests that needed so much checking for left-over clothing etc. All I knew was that the rooms looked so much nicer and we all took great pride in them.

During the 1939/45 war Mr. Morris approached Mr. Towle, regarding the future. He was worried that his craftsmen would come back from the Forces and he would not have enough work for them, so he asked Mr. Towle to place an order for furniture, for delivery as soon as possible after the war was over. I believe the timber was available and certainly the hotels had been over-used during the war and we were ready for something new. As the 1938 suites had been so successful, a deal was agreed. They were, in due course, delivered to L.M.S. Hotels all over the country, with very little alteration to the original design. But, by that time, both men had died, and their successors had to renegotiate because of the rise in the cost of living. This was a very protracted affair, not, I think, without some distress, but the companies remained good friends. We, in the Works and Domestic Services Departments always felt we would get a good job and a fair deal and, if furniture was needed, our thoughts turned to Morris automatically. We bought many chairs and tables, as well as bedroom suites and some for sitting rooms too. Neil carried on his father's traditions and John Fletcher, on our side, dealt with all the post-war problems amiably while Frank Hole, who succeeded Arthur Towle, approved the financial details. From my point of view, as one of the people who decided which rooms needed the suites and other furniture, and from that of the hotel housekeepers, it was satisfying to see the improvement in the bedrooms and the chambermaids saved labour. The day of the house porter who used to clean wardrobe tops and move furniture for the maids was coming to an end. For us all, since we took such pride in the Hotels, it was a Red Letter Day when new furniture was delivered, both in 1938 and after the war.

In 1948/9, along with the Railways, we were nationalised. Hotels of the four railway companies had to be integrated and little new furniture was ordered until we started on schemes for modernisation and additional private bathrooms. Again Morris suites, chairs, stools and tables were supplied, but of lighter wood and different design. The guidelines set down by Mr. Towle and Mr. Morris was followed, but perhaps their firm and personal attention to detail was missing. The quality still remained, but rooms were now smaller and we had to adapt to modern times. Fitted furniture became fashionable and necessary, because the rooms were smaller. White plastic replaced timber for ease of cleaning, divans replaced beds and so on. But there were always rooms in some of the Hotels with really old-fashioned furniture, dating back to the first Railway Hotels, a hundred years ago, that used up the displaced Morris Suites. Managers, Housekeepers, and even the maids knew what we meant when we said "Morris Suites" and often welcomed them as an improvement. Eventually, with replanning and the many changes, some of Mr. Morris' original furniture has been sold. It did yeoman service and has stood the test of time in most ways, setting standards of comfort and convenience for future hotel guests. Anyone who finds a piece of finely crafted Honduras mahogany furniture and sees the "Morris of Glasgow" bronze disc inside a door or a drawer might consider himself lucky. Even Arthur Negus would touch the timber, value the tiny continuous hinges and say, "That has quality." His successors in years to come probably will."   *Dorothy Crawford to Neil Morris*

Selection of Hotels and Restaurants

Gleneagles- Double and single room furnished from product lines
Cumberland hotel - beds, pedestals, bureau, dressing tables
Caledonian Hotel Edinburgh - Furnishing of single and double rooms, publicity work on hotel accommodation
Dar - es - Salaam Hotel - Zanzibar - outfitting of rooms
Royal Hotel Plymouth - supply wardrobes beds etc.
Regent Palace Hotel- general outfitting
George Hotel - cocktail bar,
Station Hotel Perth - general outfitting, beds, wardrobes, pedestals etc.
Saint Enoch Hotel Glasgow - outfitting single and double rooms
Adelphi Hotel - single and double beds,
Turnberry Hotel - outfitting single and double rooms
Euston Hotel - general outfitting supply wardrobes beds from product lines
British Railways Hotel, group product range 4180 - 12 to 15 a month.
Central Grand Hotel Khartoum - general outfitting
Strand Palace Hotel - beds, wardrobes etc.
Grosvenor House extension - product lines, beds, tables wardrobes etc.
Hospital in Belgian Congo - steel framed beds lockers wardrobes tables etc.
The Rogano, Glasgow-panelling, tables, chairs and general outfitting

The Turnberry Hotel

The Bath Hotel Glasgow

Pioneering the Helicopter

The origins of the modern helicopter and its development can be traced to the activities of Juan de la Cierva, a Spaniard. In 1926 he came to Britain and flew a converted Avro 504 biplane before an audience at the Royal Aircraft Establishment, Farnborough. Lacking wings, this Autogiro was a sort of halfway house between a helicopter and an aeroplane. Atop the fuselage was constructed a cage and above the cage was set a huge, freely rotating wing or propeller. The rotating wing or blades only worked when the wingless aeroplane went fast enough to turn them and, when they turned, lift was created and the Autogiro took off. After the demonstration the Autogiro was hailed as a success and attracted some orders for the Royal Air Force. William Weir and his brother, James were enthusiastic about the design, but what they wanted was a pure helicopter that would be cheap and easy to run, a sort of aerial Volkswagen. James formed the Cierva Company with Juan de la Cierva and held the patent rights to all of Juan de la Cierva's designs. The Cierva Company was desperately short of working capital, since few of its designs had attracted orders for the mass market. For the military, the Autogiro was a curiosity, which, when flown by them and, even though it was brand new and in its infancy, they thought had reached the end of any development.

William Weir had influence with the government, but he made it plain to James that he would not use this influence to attract any government orders. But what William Weir did in 1932 was to form an Autogiro Department at the Cathcart works of his company, J & G Weir, when he took out a licence to produce Autogiros. The Weir Company needed personnel and, from ICI, came "Doc" Bennett, who was to design the autogiro blades. The designer for the Autogiro Department was C.G.Pullin, who also designed the engines. Pullin came from Douglas Motors. Personnel who were transferred from the

Weir Company to the new department were Tom Nesbit and Ken Watson.

In April 1932 William Weir sent for Harris Morris. He had met Harry Morris through his friendship with the Paisley family of Coats and he was not afraid to make his friendships work. "Morris can you make me helicopter blades?" Of course Morris said, "Yes." The blades, to a Bennett design, were made around a single steel pipe. At intervals along its length were placed aerofoil shaped wooden bulkheads which were glued and bonded onto three ply wood and the whole blade covered with canvas.

Charlie Simm and Wattie Sutherland manufactured the first Morris blades for the first Weir Autogiro, the W1, in 1933. This first Scottish Autogiro was flown at Abbotsinch by Juan de la Cierva himself and was an aerodynamic prototype for the Avro built Cierva C.30. Morris blades were then made for the next Weir Autogiro in the series, the Weir W 2, which tested Bennett's new rotor head. Incidentally this Weir design still exists at the Royal Scottish Museum of Flight at East Fortune. Initially these Morris blades were made to a German aerofoil pattern, but later the design was changed to an American specification.

Working at Cathcart during this early period, 1934, was Otto Reder, a German, who was made responsible for single blade research. In addition to his aerodynamic experience, he was closely linked to gas-turbine (jet engine) manufacture in Germany, with such firms as Heinkel, Hirth and Junkers. (Later it was discovered he was a spy, working to find out developments with the jet engine and the helicopter in Scotland.) Reder worked closely with Harry Morris at Cowcaddens and produced a prototype blade shaped like a gigantic letter A. Though this design was not progressed with, he did install one of his smaller blades on a sailing boat of the Redwing class. This boat was built for Lord Brabazon and was tested in the Solent in 1935. Reder removed the sail and installed a rotor of eighteen feet diameter. The whole system worked perfectly, but the vessel was lost in a collision with other yachts of its class and William Weir felt that such research, though interesting, did not merit the financial outlay and the idea was shelved.

When making the rotor blades great care had to be taken when bonding the wooden parts or internal stations to the three ply wood. Morris had to develop new glues, which would bond the wood together without failing, when the wooden rotor blade rotated. Initially Harry Morris was in full control of this research work and it was to become the focus of research and development for the Morris works as Weir continued their development. At Glasgow's Royal Technical College, now Strathclyde University, all the stress calculations and all the glues for bonding were tested. The choice of blade wood was important; it had to be strong, it had to accept the constant change in temperature and pressure and it had to last. Such was the success of this work, that Weir never had any problems with the Morris blades which was in contrast to those used in Royal Air Force Autogiros. Their blades tended to warp, with the canvas and ply wood separating over the whole length of the blade.

From the W1 came the W2, W3 and W4 with Morris blades powering all four types. In the W3, Bennett came closest to developing what we would now call a helicopter. By pulling a lever, the W3 would jump-start vertically into the air and then fly horizontally. However it could not land vertically like a helicopter. The Morris involvement with the Weir Autogiro was even more in - depth with the development of the Weir W4. Morris supplied plywood, which was used for building the fuselage, together with the blades. This autogiro used all the experience of the jump-start technology of the W3 and was designed by C.G.Pullin, with "Doc" Bennett designing the rotor blades and the rotor hub. The W4 was completed at Cathcart and was readied for flight at Abbotsinch by H.A. "Alan" Marsh. Well, try as he might Alan Marsh could not get the W4 to fly. He revved up the engine, he clutched and de-clutched the rotor but still the W4 would not take off.

At the airfield the audience included H. Morris, William Weir, James Weir, C.G. Pullin, "Doc" Bennett, Ken Watson and T. Nesbit. Everyone decided that the problem could be solved after lunch. Then the same audience gathered again. Alan Marsh tried again the same as before; revving, clutching, de clutching - nothing happened. Then "Doc" Bennett and Ken Watson mounted the W. 4 and adjusted the rotors and the rotor hub, with Alan Marsh still in the cockpit. Marsh revved up the rotors and turned the engine to its maximum power, clutching the rotor mechanism. Suddenly, the W 4 fell on its side to the scream of the over - revved engine. The W 4 was smashed to pieces and the broken blades flew in all directions with the audience fleeing to howls of laughter. Luckily, Alan Marsh only suffered cuts and bruises and managed to pull himself clear of the wreckage. No one ever discovered why the W 4 could not take to the air for, by 1936 the Autogiro was clearly obsolete. In Germany Henrich Focke had developed a workable helicopter, with side-by-side rotors, called the Focke Achgelis Fa 61. It had been test flown inside the Berlin Exhibition hall by the aviatrix Hanna Reitsch after the Berlin Olympic Games. The Germans proved to the world that the helicopter was a viable proposition and clearly showed that the Autogiro was dead. The remains of the W 4 were resurrected as a rotorcraft with superimposed Morris blades and an aerodynamic fuselage, the “Jabberwock.”. Clearly the design would never work and Morris calculated and suggested to Weirs that the side by side rotors, as in Henrich Focke’s machine, were the best layout for the first British helicopter, the Weir W5. The rotors of the W 5 were held to the fuselage by special plywood box structures that could accommodate the strain and the power of the single W 5 engine. In the late nineteen thirties the W 5 first flew at Dalrymple in Ayrshire. Such was its success that the Air Minister, Lord Swinton, clothed the helicopter trials in a veil of secrecy.

The next design in the Weir series was the W 6. This was a much more powerful helicopter than the earlier W.5 and had the facility for carrying two passengers. Pullin recognized that the early Autogiro blades would not be strong enough to fly such a powerful machine and asked "Doc" Bennett to design a new blade. This blade incorporated much stronger wood, a metal leading edge and was the first to employ a new bonding process, Hydrolon, pioneered by Neil Morris, who had taken over from his father and the Royal Tech. When the W.6 was first flown it could do everything and more than a helicopter could do up to that time, including flying with three people, fly backwards, forwards and even fly vertically. One test at Thornliebank involved the W.6 flying in a figure of eight over a closed course and then hovering above a plywood panel, touching it occasionally. Such was the strength of the blades that when the W 6 was flown in very windy weather it was not the blade that broke but its linkages and drives. Neil Morris had voiced an opinion that this could happen when he had strengthened the blade by adding a sliver of Canadian Spruce down the blade length, making it stronger. This was a technique that had been used for the Hafner Autogiro, flown by A.E.Clouston and financed by Coats in 1938. The wooden blades were now stronger than the propeller hubs.

By now the war had broken out and in, May 1940 J & G. Weir decided to pull out of the helicopter field, with the Weir series of helicopters ending. Weirs had not been paid for their work on the W5 and W6, or for supplying the Royal Air Force with a Squadron of Cierva rotorcraft, which were used for radar calibration and army cooperation work. Weir had constantly asked the government for payment, but no payment from the government was forthcoming. Weir then placed the W5 and the W6 into storage at their works, later moving the airframes back to James Weir's home at Dalrymple in Ayrshire.

One field of research that the Glasgow firm of Weir was involved in, was the development of Frank Whittle's jet engine. William Weir had suggested to Pullin and Bennett that blowing pressurised air through a rotor blade with a gas turbine could take away rotor vibration and ensure trouble-free flying. (This was the idea behind the W.8.) The piston engines of those days were causing severe airframe and rotor vibration when the helicopter was flying. Vibration would increase and travel through the airframe and then to the wooden blades, destroying stability. Weir believed that the firm of Power Jets, which developed Whittle's engines, would be able to help. So, when Mark Bonham - Carter, one of the directors of Power Jets, asked Weir to help finance the first jet engine for a British aeroplane, William Weir put up much of the capital for the project. By an arrangement with the new Ministry of Aircraft Production, many of the Weir technical team went to Power Jets, including C.G. Pullin and Ken Watson. This arrangement seemed to work well, for, by 1941 Britain had a jet in the air. The success of the engine led to a conflict between G. & J. Weir and the Minister for Aircraft Production, Sir Stafford Cripps. Politics entered into the scene and Cripps told Weirs that their shares in Power Jets were to be nationalised and to be owned by the state. The idea for nationalisation had come not from Stafford Cripps but from Frank Whittle himself, the inventor of the jet engine! Though the compensation was generous, the Weir brothers were not happy about the arrangement and when they were finally denied direct access to Power Jets to see the fruits of their investment, James Weir felt it was the last straw. Using the Power Jet compensation Weir’s reformed the Cierva Company in the autumn of 1943. Once again they sent for Morris and asked for two things, firstly a set of helicopter blades and secondly a moulded fuselage for an experimental helicopter, by way of compensation, the new Cierva

Company had received an order for an experimental helicopter, to be tested by the Air Ministry in 1943. The new helicopter was to be the Cierva W 9 and was designed to have a moulded wooden fuselage built around a tubular steel airframe.

By this time in the war Morris had built up a reputation for working out problems with moulded wood and aircraft plywood. Neil Morris had worked with the Royal Aircraft Establishment and Ministry of Aircraft Production in developing aircraft plywood and was instrumental in curing the d Havilland Mosquito problems by the use of the autoclave and new bonded woods and wood glues. For the W 9 a massive autoclave was built at Cowcaddens. It represented a lot of problems for Morris due to distortion of the wood under pressure and warping when the fuselage was removed from the autoclave. It took nearly a year to solve the problem and, when the fuselage of the W 9 was sent to Cierva, the helicopter was finally completed. When the W 9 took to the air for the first time in October 1944 it handled badly in the air due to its Autogiro rotor head and crashed. When it hit the ground Ken Watson nearly lost his life when his head was pushed through the cockpit into the path of the whirling rotor, his quickness saving his life. The Cierva W9 used the engine and a lot of the components of the Weir W6, which had been in storage at Dalrymple in Ayrshire.

The Cierva W 9 was rebuilt as the Cierva W.9a which was a much more successful helicopter and flew with Morris blades during March 1945. In July 1944, Neil Morris had been present at the intelligence de-brief of Louis Breguet in France. Breguet had been involved with the Germans in manufacturing helicopters, including the giant Focke Achgelis FA284, which was to be built just outside Paris. The details of these blades caused a change in the way Morris designed their helicopter blades. The first sets of blades for the W.9a were identical to the W 6 and were much stronger than any blades then in production. The next set of blades incorporated all the skill of the

Morris Factory; they were made of moulded wood, but followed the German practice as described by Louis Breguet to Neil Morris. They were built just like a wing with internal stations and were bonded together in an autoclave. Charlie Simm was in charge of this building work, while Neil Morris oversaw the completed blade. Once again Glasgow's Royal Technical College was involved in the research and development with the new blades being a complete success. The expertise in blade design had been recognized when the deputy director of the Airborne Forces Experimental Establishment (A.F.E.E.), J.A.J. "Doc." Bennett ordered eight sets of blades for an experimental Autogiro. Bennett was the first director of the A.F.E.E. and had been appointed in August 1940. When he left for America in 1941 another director was appointed. He was none other than Raoul Hafner, the early helicopter pioneer. He had designed the Rotachute (the blades had been ordered by Bennett in 1940) for the paratroops. The Rotachute was intended as a simple Autogiro, which could hold one soldier and a Bren gun and was to be carried in a Wellington bomber, which had had its tail turret removed. The Rotachute would be dropped from the turret and then be controlled and guided through a huge rubber mounting under the rotor by the paratrooper.

It was envisaged that, in combat the Rotachute would land immediately at the designated target. Sets of blades were sent from Glasgow to the establishment at Ringway, then Sherbourne - in - Elmet and later, Beaulieu, where about a dozen Rotachutes were built. There they were towed into the air by a Tiger Moth or a Bentley motorcar and appeared very practical. They were, in some ways, similar to the FA 330 Autogiro used in German submarines. Though they could possibly have been a success, the whole concept was impractical. The design went to America with the end of the war, as did the design of the hypersonic Miles jet fighter. The Malcolm Company also built a flying jeep, which used Morris blades. These were truly the longest blades ever made at Cowcaddens. The airborne forces had a requirement for 138 of these flying jeeps. General Eisenhower was also pressing the development, for the Rhine crossing. The wooden blades were of a new pattern, designed by Raoul Hafner, which were mounted above a Willys Jeep. The blade was moulded onto a hollow socket and termed the R. 46 blade. Initially the jeep got into the air, when towed by a Bentley motorcar and then, later in the test phase, it was towed by a Whitley glider tug which took it up to about two thousand feet. In the air Squadron Leader Little, the pilot, found the whole contraption difficult to fly and control. Even the shortest flight was very tiring. On one occasion flying was so difficult Little collapsed after the flight.

Though brilliant ideas, the Rotachute and the towed jeep were not practical propositions and never progressed from the experimental stage. At one point, during a test phase at Sherbourne-In-Elmet, none other than the jet engine pioneer Frank Whittle put in some advice on the flying jeep project, but Mr. Neil told him that his advice was not needed. Early in 1945, Cierva ordered three sets of Autogiro blades for the war - surplus Avro Rota Autogiros that they had bought from the Air Ministry, thus continuing the Morris connection with the pre - war Autogiro.

At the end of the war, in May 1945, the Ministry of Aircraft Production (MAP) sent many of its staff into Germany and the former occupied Europe, to inspect German aircraft development. Since they were entering a combat zone the staff were put into uniform and given temporary rank. One such officer was Flight Lieutenant Neil B. Morris. He was sent to inspect German helicopter development. One example of the Focke Achgelis Drache, captured near Berlin, was selected to fly to Britain. As it crossed the English Channel to be flown to a designated base, Flight Lieutenant Morris became the first passenger to cross the English Channel in a helicopter. This Drache crashed in an accident, but it gave an insight into how the helicopter could be developed and much of the test data was not lost to the aircraft industry.

The German rotor - blades, using an American aerofoil pattern, as described to the Allies by Louis Breguet, were felt to be superior to the blades then being made. At the Cierva Company, Pullen and his assistant, Jacob Shapiro were working on a new giant helicopter called the "Air Horse". In reality, it was really an update of the Weir W6 and utilized all the wartime German giant Focke Achgelis technology. It was nicknamed by the press the, "Spraying Mantis". It was designed as a three-rotor helicopter, to provide crop spraying for the fields of Britain's East African Empire. With the Ground Nut Scheme in full swing, pest control by helicopter seemed the natural way to advance insect destruction. Powered by a single Merlin engine, the Air Horse took to the air in the late forties. It was the biggest helicopter that had flown up to that time and the wooden rotor - blades were a marvel of the woodworker's skill. The blades were not straight, but twisted with a four - degree bend and, as Neil Morris constantly pointed out to Shapiro, the blades were immensely strong. They were in fact stronger than the hubs. The blades were made in sets of three and used special aircraft techniques in bonding and curing, which Morris had pioneered during the war. Each of the blades was coloured in relation to its position on the helicopter. At the nose was a three bladed rotor above the cockpit and at the opening tail, mounted on huge aerofoil outriggers, were the other two rotors, making nine blades for each helicopter. Morris manufactured and repaired at least four sets of blades for the Air Horse.

Two of these giant helicopters were built and they seemed to be what was wanted at the time. But there was a feeling of uneasiness at Morris, concerning the strength of the helicopter hubs and as a quick test of strength and construction a complete set of blades was sawn up and tested by Cierva at Eastleigh near Southampton. In one blade an internal bulkhead/ station had been moulded onto the outer contour of the blade at an angle, but Shapiro had to agree that its construction would not affect the helicopter's handling. To strengthen the blade further Mr. Neil added strips of Canadian Birch down the length, as in the W 6 of 1938. This was at the request of Shapiro, who by now, was in charge of the complete Air Horse programme.

1950 dawned a new era in the British aviation industry, with Cierva at last seeming to be on the verge of success. On 12th June 1950 Alan Marsh completed a series of trials with the Air Horse, including emergency landing, hovering, and autorotation at Southampton. The next day, a Tuesday, seemed ideal for flying the Air Horse. She was rolled out, fuelled and the next test phase entered into - hovering and descending. By four o' clock Alan Marsh felt that flying should end that day and was ready to leave Eastleigh Airport and go home. Unexpectedly, "Jeep" Cable arrived at the airfield and asked if he could fly the Air Horse one more time, before the end of the flying that day, which was teatime. Much against his better judgment, Marsh agreed to the flight and ordered the giant helicopter to be fuelled up and prepared for another flight. "Jeep" Cable was the helicopter test pilot for the Ministry of Supply and had been selected to train other pilots on the Air Horse. Unusually, Cable was the only pilot in the Royal Air Force who had never flown a fixed wing aircraft. When he left the service he had gone to the Ministry as their senior helicopter test pilot. Cable was very anxious to get as much flying time in on the type as possible to get more experience. Giving him control, Marsh allowed Cable to fly the Air Horse at two hundred feet around the locality. Suddenly, as they were flying over a copse, the aircraft nose went down, pitched in at an angle and hit the ground. The Air Horse exploded in a ball of flame killing all three of the occupants. All that was left was the tail and, surprisingly, the rotors. During "Jeep" Cable's gentle manoeuvres the nose rotor hub had failed, breaking a linkage and shedding a blade. This blade took off the other two and all stability was lost with the helicopter crashing.

Many were the sleepless nights for Morris of Glasgow, before the court of inquiry published its initial findings. They found the crew had died by misadventure and that the fault in the helicopter was at the hubs and not the blades - Morris was exonerated. With the loss of the Air Horse, J. & G. Weir lost all interest in helicopter development and sold the Cierva Company to Saunders - Roe in January 1951. The Saunders - Roe company amalgamated with the helicopter divisions of Fairey and Bristol aircraft and, ultimately, with Westland Helicopters. Some of the Cierva personnel moved to Saunders Roe with the amalgamation.

By April 1951 Morris of Glasgow decided to look again at its helicopter blade manufacturing capability. They had supplied the blades for all of the British helicopters up to that time including sets for the Bristol 171 (Sycamore) and the Bristol 173 (Later the Belvedere.) These blades were based on the ones used by the flying jeep, to the design of Raoul Hafner. They had also supplied blades to Louis Breguet in France who was developing a series of twin rotor helicopters based on German design of the Focke Achgelis Drache. H. Morris & Co. felt that continuing with this enterprise did not justify the amount of investment, capital and time and, sadly, the company left the field of developing the helicopter blade, to concentrate on newer projects.

An early autogiro over the countryside

A military Cierva 8L shows its paces. Clearly seen are the rotating blades and Avro 504N fuselage

The first Weir autogiro the Weir W1

The Weir W-2 over Hanworth

Alan Marsh with Juan de la Cierva in the W3

An Avro built Cierva C30.Tthis was the first really successful autogiro which was produced in quantity in France, Germany and America as well as in Britain

The wooden mock– up of the Weir W4

Dubbed the Jabberwock here projected in wood with twin vertical rotors, this helicopter was never completed

The unsuccessful Weir W4, which crashed on take – off

The first British helicopter, the Weir W5, with rotors on the outriggers

The first flight of a British helicopter, the Weir W5 at Thornliebank

The Weir design team, which includes M. Brennan (centre) with the Weir W5. Brennan went to Saunders – Roe to design rockets

The Weir W6 under tow on its way from Thornliebank to Dalrymple

The Weir W6 under construction at Thornliebank. It followed closely the design of the first German helicopter, the Focke Achgelis, flown by Hanna Reitsch in 1937

The Weir W6 with James Weir at Dalrymple in Ayrshire

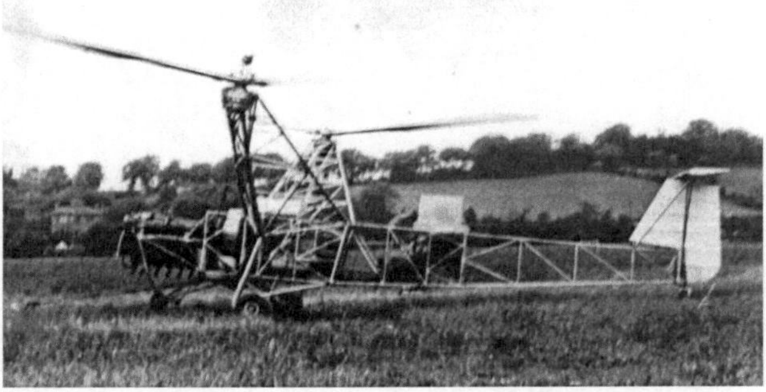

First run of the W6 before its first flight

The Weir W6 in the air at Dalrymple, Raymond Pullin, pilot.

The blades of the Weir W6 in various stages of construction at Milton Street

At Thornliebank in 1939 the W6 helicopter with A.V.M. Tedder as passenger

An early Hafner design in the air, the AR III flown by A E Clouston

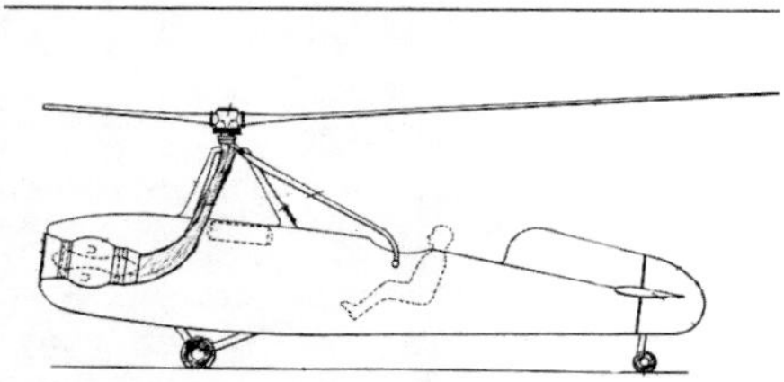

The Weir W8, which was to use hot air blown through the Morris blades by an early jet engine

Hafner's first autogiro financed by the Coats family 1933

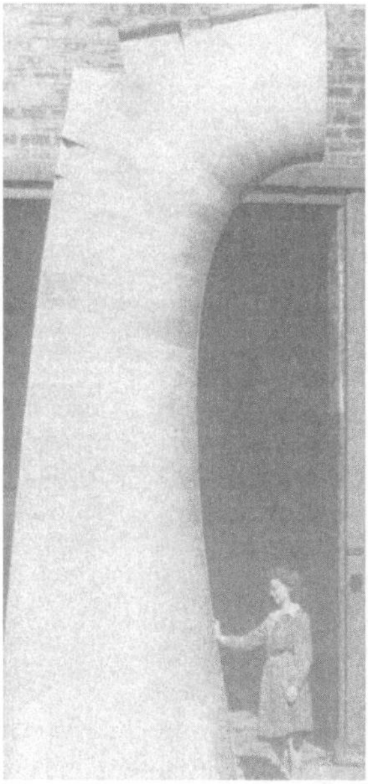

Cierva W9. It caused Morris a lot of problems due to the size of the autoclave. Completed in 1944 it awaits its first flight

The modified Cierva W9a in the air at Southampton

After its accident the W9 received a new fuselage

Proving the strength of the Weir W6 blades at Milton Street

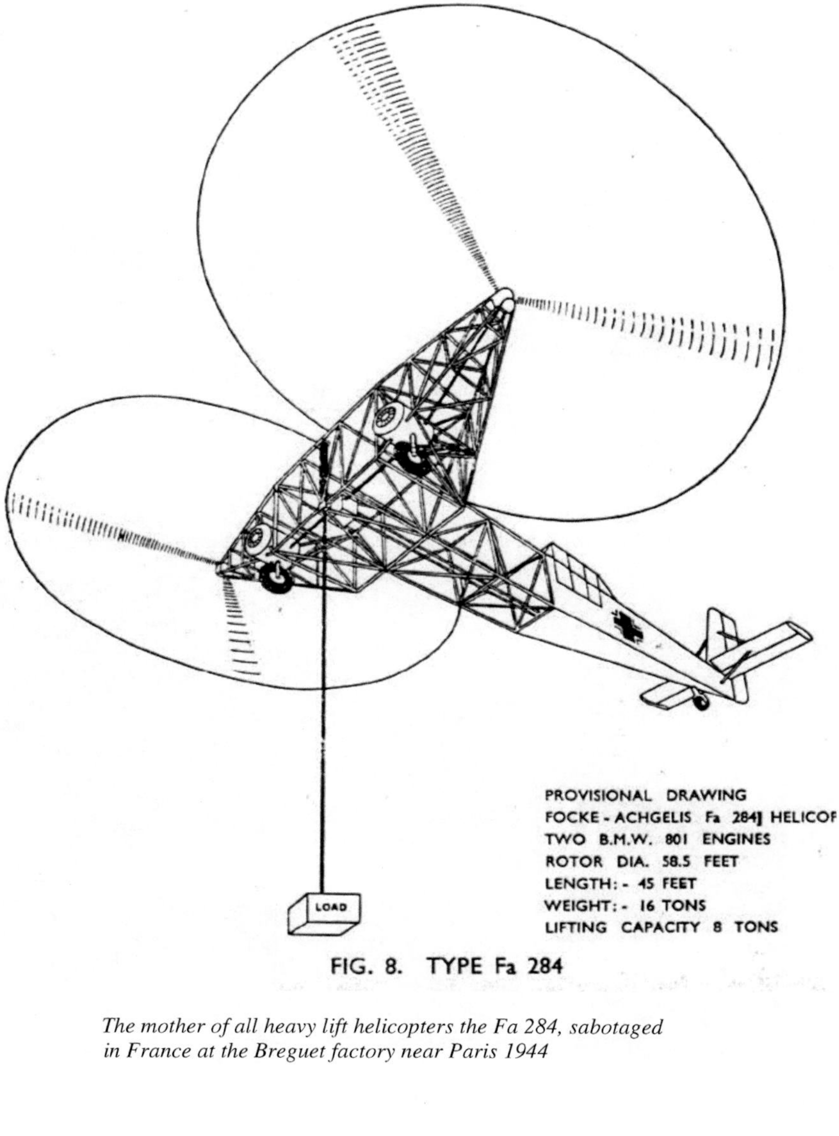

FIG. 8. TYPE Fa 284

The mother of all heavy lift helicopters the Fa 284, sabotaged in France at the Breguet factory near Paris 1944

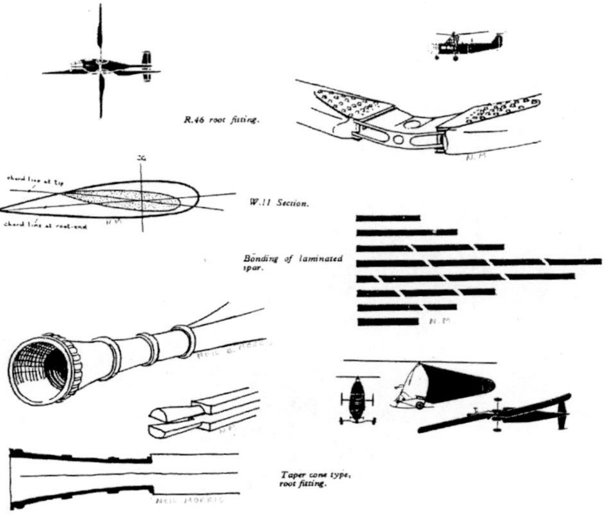

A selection of blades designed by Neil Morris with designs by Hafner/Bennett and Breguet

An Air Horse blade awaiting delivery at Milton Street

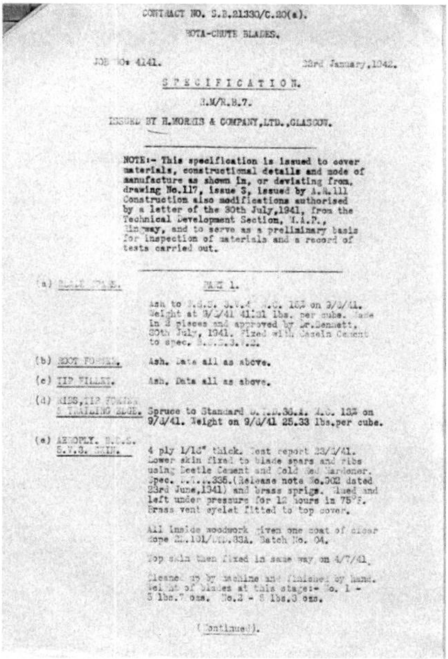

CONTRACT NO. S.B.21330/C.20(a).

ROTA-CHUTE BLADES.

JOB No. 4141. 23rd January, 1942.

S P E C I F I C A T I O N.

R.M/R.B.7.

ISSUED BY H.MORRIS & COMPANY, LTD., GLASGOW.

NOTE:- This specification is issued to cover materials, constructional details and mode of manufacture as shown in, or deviating from, drawing No.117, issue 3, issued by A.R.111 Construction also modifications authorised by a letter of the 30th July, 1941, from the Technical Development Section, M.A.P., Kingsway, and to serve as a preliminary basis for inspection of materials and a record of tests carried out.

PART 1.

(a) [illegible]. Ash to [illegible] on 3/3/41. Weight at 3/3/41 41.31 lbs. per cube. Made in 2 pieces and approved by Dr.Bennett, 30th July, 1941. Fixed with Casein Cement to spec. [illegible].

(b) ROOT FORMER. Ash. Data all as above.

(c) TIP FILLET. Ash. Data all as above.

(d) RIBS, TIP FORMER & TRAILING EDGE. Spruce to Standard [illegible] 13% on 9/3/41. Weight on 9/3/41 25.33 lbs. per cube.

(e) AEROPLY. [illegible] SKIN. 4 ply 1/16" thick. Test report 23/3/41. Lower skin fixed to blade spars and ribs using Beetle Cement and Cold Red Hardener. Spec. [illegible] 335. (Release note No.302 dated 23rd June, 1341) and brass sprigs. Glued and left under pressure for 12 hours in 75°F. Brass vent eyelet fitted to top cover.

All inside woodwork given one coat of clear dope [illegible].33A. Batch No. 04.

Top skin then fixed in same way on 4/7/41.

Cleaned up by machine and finished by hand. Weight of blades at this stage:- No. 1 - 3 lbs.7 ozs. No.2 - 3 lbs.3 ozs.

(Continued).

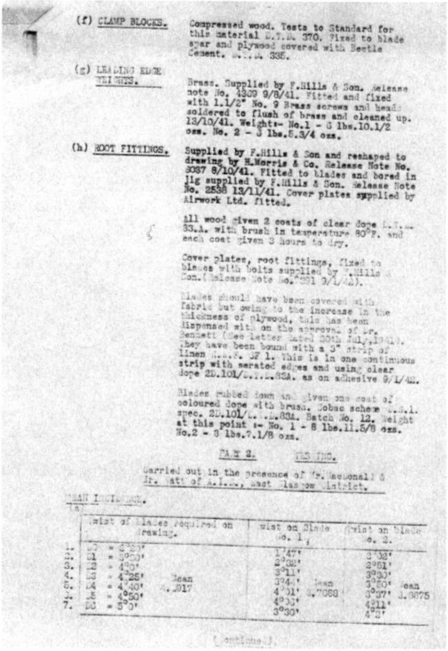

(f) CLAMP BLOCKS. Compressed wood. Tests to Standard for this material D.T.D. 370. Fixed to blade spar and plywood covered with Beetle Cement. [illegible] 335.

(g) LEADING EDGE WEIGHTS. Brass. Supplied by F.Hills & Son. Release note No. 4309 9/8/41. Fitted and fixed with 1.1/2" No. 9 Brass screws and heads soldered to flush of brass and cleaned up. 13/10/41. Weights:- No.1 - 3 lbs.10.1/2 ozs. No. 2 - 3 lbs.5.3/4 ozs.

(h) ROOT FITTINGS. Supplied by F.Hills & Son and reshaped to drawing by H.Morris & Co. Release Note No. 3037 8/10/41. Fitted to blades and bored in jig supplied by F.Hills & Son. Release Note No. 2538 13/11/41. Cover plates supplied by Airwork Ltd. fitted.

All wood given 2 coats of clear dope [illegible] 33.A. with brush in temperature 80°F. and each coat given 3 hours to dry.

Cover plates, root fittings, fixed to blades with bolts supplied by F.Hills & Son. (Release Note No. [illegible] 9/1/42).

Blades should have been covered with fabric but owing to the increase in the thickness of plywood, this has been dispensed with on the approval of Dr. Bennett (See letter dated 30th July, 1941). They have been bound with a 3" strip of linen [illegible] 1. This is in one continuous strip with serated edges and using clear dope 2D.101/[illegible].83A. as on adhesive 9/1/42.

Blades rubbed down and given one coat of coloured dope with brush. Cobac scheme [illegible] spec. 2D.101/[illegible].83A. Batch No. 12. Weight at this point :- No. 1 - 8 lbs.11.5/8 ozs. No.2 - 8 lbs.7.1/8 ozs.

PART 2. TESTING.

Carried out in the presence of Mr. Macdonald & Mr. Watt of A.I.D., East Glasgow District.

MEAN INCIDENCE (?)

	Twist of blades required on drawing.		Twist on Blade No. 1.		Twist on blade No. 2.	
1.	[illegible] = [illegible]		1°47'		2°08'	
2.	D1 = 3°00'		2°32'		2°51'	
3.	D2 = 4°0'		3°11'		3°30'	
4.	D3 = 4°25'	Mean	3°44'	Mean	3°50'	Mean
5.	D4 = 4°40'	4.2917 (?)	4°01'	3.7068	3°37'	3.8675 (?)
6.	D5 = 4°50'		4°03'		4°11'	
7.	D6 = 5°0'		3°30'		4°2'	

(Continued).

The specification of the Rotachute blade

The one man Rotachute prototypes

The Rotachute on test on the back of a former balloon tug lorry

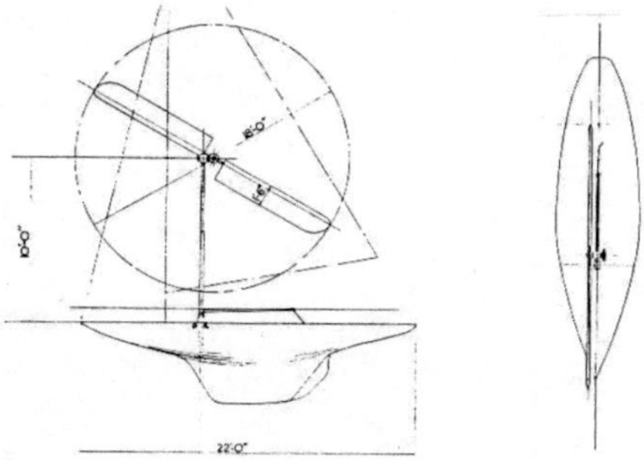

Otto Reder's yacht

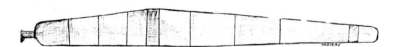

The new helicopter blade, a British copy of a German design

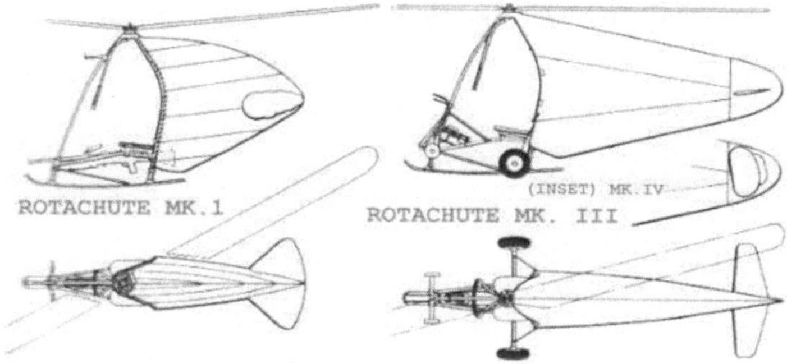

The Rotachute in various flight patterns

The R46 blade for the Flying Jeep displayed on the roof at Milton Street.

The Flying Jeep or Rotabuggy

Rotabuggy or Flying Jeep in the air

Marcel Lobelle with the flying jeep, Sqdn.Ldr Little in the cabin

The first real practical helicopter, the FA223, which flew the channel in 1945 with Neil Morris on board as an observer

The FA223 Drache (Kite) in Luftwaffe service

A ground view of the Air Horse, notice the huge blades earning the name, "Spraying Mantis"

A wooden model of the Air Horse

A contemporary view of the Air Horse taken from a comic

A ground to air view of the air Horse. It is difficult to imagine that it was powered by the same engine as the Spitfire- one Rolls Royce Merlin.

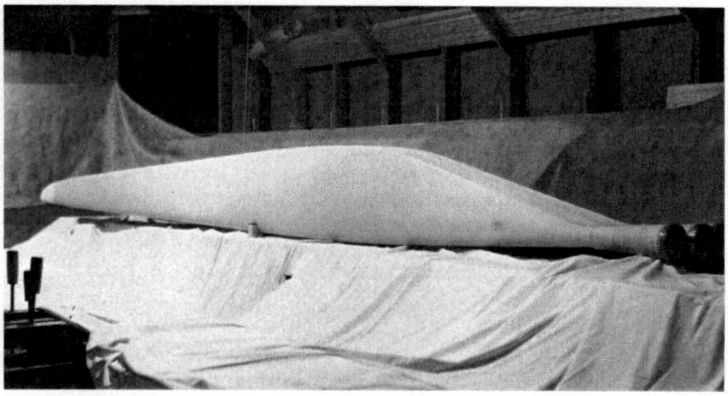

A finished blade at Milton Street

An offside view of the Cierva Air Horse on the ground, the door was always open to help the engine to cool.

The crash of the Air Horse

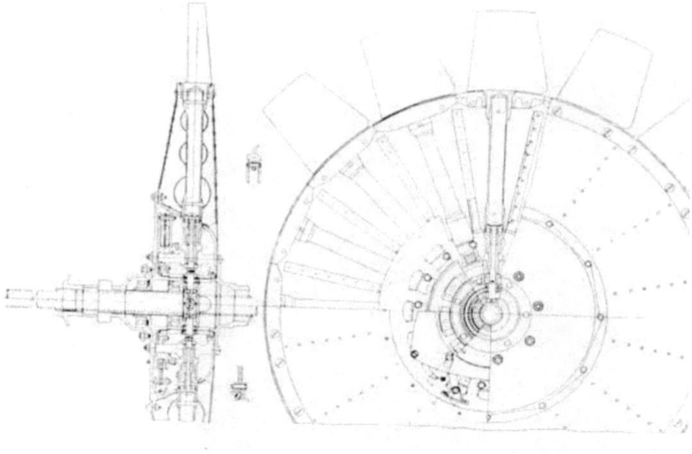

The cooling fan of the Air Horse designed by Neil Morris

The Bristol Sycamore on the ground, the blade design taken from the Flying Jeep

A Royal Air Force Bristol Sycamore in service for air sea rescue.

Sycamore blades awaiting bonding at Milton Street.

The much larger Bristol 173 powered by Morris blades

The Flettner Helicopter inspected by Neil Morris for the Ministry of Aircraft Production

Morris were consulted by the Royal Navy on blade design for the first Sikorsky helicopters in naval service.

Sweeping the Sawdust

During the Great War most of the aeroplanes then in production were made out of wood, but, because of wood shortages towards the end of the war, the Ministry of Munitions decided that all aeroplanes were to be made out of metal. Even when the Air Ministry was formed in 1918, they continued the change in policy, regarding wood as an obsolete aircraft material and that all - metal aircraft offered the Royal Air Force better serviceability and easier construction. By 1937 the international situation was clearly taking a turn for the worse and Britain had to re – arm, quickly.

One day at Cowcaddens, “The hunting party” arrived. The “hunters” consisted of Lord Swinton, the Air Minister, William Weir, James Lithgow of the Greenock shipbuilding concern, Lithgow Shipbuilders, together with C.A. Oakley and others, the party having been organised by William Weir. The hunting party was none other than a sub-committee of the Committee of Imperial Defence. They were visiting factories around Glasgow to see their wartime production potential and the party had been organised by William Weir as a security cover. Harry Morris showed them around the factory at Cowcaddens and the work that was being done for the Weir helicopters and as a result of this visit the committee recommended that H. Morris & Co should be put on the list of approved suppliers of aircraft wood for the Air Ministry. One of the first supplies of aircraft wood was for Scottish Aviation Ltd which was to go on to manufacture aerial targets (Queen Bee) for the Royal Air Force. Such was the quality of the wood and plywood panels, that most of the aircraft firms used Morris as a supplier of aircraft wood. This also meant that Morris received one - off contracts to supply wood for such establishments as Farnborough and the Marine Aircraft Experimental Establishment at Helensburgh. The Committee of Imperial Defence had also decided that all training aircraft were to be made of wood and plywood, to preserve strategic materials such as aluminium.

Timber Control, of the Board of Trade, had been installed at Park Circus in Glasgow, at the site of an old hospital. A further branch of Timber Control was set up to deal with all plywood matters and that was the Plywood Branch. The two parts of the Plywood Branch had clearly defined roles; one part would deal with construction or general-purpose materials, while the other branch dealt with the services, particularly the demands for aircraft manufacture. Planning of this department had begun well before the war years, with Canada being seen as the prime producer of plywood, particularly Douglas fir and birch. Due to the demands of the aircraft industry a further department was set within Timber Control and that was the Technical Section. For H Morris & Co the existence of the Technical Section meant that the company was relieved of buying plywood and disposing of the finished product. But it also left the company the capacity to develop the specification of aircraft plywood.

Up to December 1939 plywood for stressed aircraft parts was made to BSS 4V3 and this demanded a high quality of veneer. A new specification was issued for aircraft plywood BSS 5V3 and this was issued at the turn of 1939. To the surprise of the industry Morris took on all research and development of 5V3 plywood, which they called, "Aeroply." By January 1941 5V3 was being produced at 250 boards a day by 50 people amounting to 7000 sq. /ft. a day.

New adhesives were developed to bond the new plywood. Phenol-formaldehyde - in film or liquid form - was used for some varieties of plywood, while casein glue was used for others. When the panels were glued up they were pressed after fifteen minutes at 200lb per sq.in. Birch was cut on rotating machines and cut with great accuracy. Other woods were used and investigated; Spruce, Western White Birch, Canadian Yellow Birch, Rock Maple, European Birch and Beech.

In addition to the Mosquito using 5V3 plywood, tactical gliders such as the Horsa and the Hamilcar were largely plywood built. For the Ministry of Supply, Morris supplied "Bridging" plywood for the Bailey bridge pontoon. Plywood was also used for Landing Craft and the Admiralty demanded plywood for use in the Fairmile system of prefabrication, for the decks and deckhouses of M.T.Bs and other light naval craft. The M.T.Bs were also built up on frames of Morris plywood. Veneers were glue bonded and compressed and used for many purposes such as propeller hubs and bearings, bomb doors, noses such as that for the Liberator transport (C87). Camera and Perspex framing were made from veneers glued together to give the required thickness. They were also found to be immensely strong. For research and development, Morris made compressed wood sections of aircraft that were in large scale production, such as the Vickers Wellington and tested them to see if the concept of bonded wood replacement was viable in the light of shortages. As part of the industry, Morris was involved in the test results for the bonded Spitfire. In the event of a loss of production, a bonded Spitfire was manufactured. Though its strength was 70% of the original type, production did not take place, since the expected catastrophe did not arrive and the fuselage went on for further testing between 1940 and 1942.

The aircraft manufacturer Phillips and Powis had been purchased by Rolls - Royce and were producing a new all wood, two seat fighter trainer for the Royal Air Force. This aircraft, powered by the Rolls - Royce Kestrel, was the Miles Master and its performance was almost equal to that of the early Spitfires and Hurricanes. F. G. Miles worked closely with Harry Morris on this first major contract for a wooden military, training, aeroplane and no faults were found with any of the plywood panels or spars. Another Phillips and Powis training aircraft was the twin seat Magister which was used as a primary trainer before aircrews took to the Master. An all-wood aircraft, it was in the same class as the Tiger Moth, but it was a monoplane whereas the Tiger Moth was a biplane.

Sydney Camm, the chief designer at Hawker Aircraft, had designed the new Hurricane monoplane fighter, which was widely sub contracted throughout the aircraft industry. Around its steel tubular airframe wooden stringers, formers and longerons were placed and, over this, was wrapped a layer of fabric. This simple production technique meant that thousands of Hurricanes could be mass-produced throughout Britain. The firms of Gloster and the car company, Austin, also made Hurricanes. Morris supplied both companies with all their Hurricane aircraft timber and plywood in the period 1940-41. Gloster Aircraft appreciated greatly the quality and finish of the aircraft plywood and wrote a letter of appreciation to Harry Morris. During this same period all the plywood supplied to Avro for their twin-engine Anson aircraft, came from Cowcaddens. The Anson was a military trainer and general reconnaissance type for Coastal Command. (The design was based on an early pre-war Fokker airliner.) The fuselage was composed of a tubular steel frame, fabric covered, but the rest of the aircraft was wooden, such as the wings and floor.

The Bristol Beaufighter had been selected by Lord Beaverbrook, Minister for Aircraft Production, in 1940, as a priority to counter the threat of German night bombers. Carrying radar, this twin-engine fighter was a descendant of the earlier Blenheim bomber and was equipped with four cannon and six machine-guns. Although the Beaufighter was all metal, Morris supplied plywood for the floor, the observer's seat, the rear hatch structure and the aerial. To save weight, the cannon bay doors were made of a lattice of double - curved, alternatively paired, thin strips of wood. Another weight saving move on the Beaufighter was to make the tail tips and all the entrances and hatches out of wood. Special machinery had to be brought in for the Beaufighter contract. Harry Morris saw this as very unproductive; the machinery was used to form the wood into special shapes to fit the aircraft airframe and was not suitable for other use in the factory. Geoffrey de Havilland, founder of the de Havilland Aircraft Company, had built a four-engine airliner for Imperial Airways called the "Albatross". Termed "Frobisher Class" in Imperial Airways service, it was of all wood construction and was made in two parts in a gigantic mould. This use of wood on such a scale was pioneering and de Havilland had entered into a contract with Aero - Products Ltd., to supply him with all the materials and know - how to build the all - wood Frobishers. The aircraft took to the air just before the outbreak of the war, but the second prototype broke in two behind the main – spar, on the ground during overload tests, splitting the airframe cleanly in two. The wood had absorbed great quantities of water and started to rot underneath the plywood skin and had started to curl and break. It was not uncommon for an engineer to put his hand through the aircraft skin and pull out handfuls of wood and demonstrated, by squeezing the pulp, how water had been absorbed.

By now war had broken out and de Havilland had designed and built the all wood Mosquito as a private venture. It had only been built under the agreement of Sir Wilfred Freeman of the Air Ministry/Ministry of Aircraft Production, on the understanding that it would not use any strategic materials and even then it was for photo - reconnaissance use only. The early Mosquito developed the same trouble as the Frobisher. The Mosquito absorbed huge amounts of water in its wooden structure. At the insistence of the Minister of Aircraft Production, Lord Beaverbrook, the de Havilland problem was placed in the hands of the woodworking industry. According to Morris tradition, Geoffrey de Havilland had been on the maiden voyage of the "Queen Mary" and had noticed that the liner's doors had been manufactured to a different process than he had seen on other ships. The doors had been made to a Morris process of bonded balsa wood called "Stressed Skin Wood Construction." When he asked the builder's representative who had made the doors, de Havilland was told - "Morris of Glasgow." When he came to Glasgow he was fully aware of the type of work that was being undertaken at Cowcaddens with aircraft plywood. (The wood was patented as GB427500)

Using autoclaves and wood glues developed for the Weir helicopters, Morris cured the Mosquito problem. The fuselage bonding was the greatest source of concern and an autoclave was built to manufacture a rear fuselage for the Royal Aircraft Establishment, Farnborough which was testing the strength of the fuselage structure. The production of the Mosquito fuselage was a master of the woodworker's art. It was made in two shells along a huge mould, split along the fore and aft line. All the electrics, hydraulic lines were installed in one side, and then the fuselage was joined together. After mating the bomb doors were cut out from the fuselage before final assembly took place. The wing was made as a complete unit of plywood, balsa and birch ply, and then was fed onto the completed Mosquito fuselage.

One day a Mosquito landed at Farnborough and broke cleanly behind the trailing edge of the wing. To increase the fuselage strength a new autoclave process replaced the balsa sandwich with man-made plastic. The autoclave fuselage worked, but it was felt it would be uneconomic to introduce this process to factories making Mosquitoes because it would disrupt the production line. Neil Morris also pointed out that the stress in the fuselage was in some ways similar to that of a rotating helicopter blade. There were bending forces, attacks by heat and light and dampness was everywhere. He suggested that furniture wood glues be introduced throughout the Mosquito programme, that better kiln drying be supervised and that on the starboard side of the Mosquito a wooden spruce longeron be placed from behind the wing trailing edge to half - way down the fuselage side. It worked. De Havilland also introduced the anti - shimmy tail wheel, which had a large groove cut down the middle to stop vibration from the runway travelling to the fuselage and breaking the structure. One of the most famous signs ever seen must have been at one of the contractors, which said, "Whatever You Do Don't Spare the Glue."

The "Wooden Wonder" became the star of the Royal Air Force, flying in all versions, bomber, fighter and photo reconnaissance. The Ministry of Aircraft Production also asked Morris to investigate substituting balsa with foam in the balsa plywood sandwich panels for Mosquito plywood. When tested in October 1942 there was a drop of density in the new panels of 30% in comparison with 100% strength in the original panels. In all these roles Morris supplied know - how for new wing spars and the overload wing tanks for the Mosquito photo - reconnaissance versions. These fuel tanks, made up from beautiful laminations and supervised by Jack Cantle, were bonded and cured in small autoclaves, then sent to de Havillands. During 1943 the Ministry of Aircraft Production dropped the humidity test for Mosquito wood on the grounds of speed and economy and in so doing guaranteed the type's failure in the Middle - East and South East Asia. In these regions the Mosquito developed the old problems of the "Frobisher" and the type was withdrawn, to be replaced by the Beaufighter.

Always there had to be considered the shortage of aircraft wood. As a substitute for the balsa wood in the Mosquito plywood, the Air Ministry found timber called "Quipo". This wood was found in South America in Panama around the Isthmus of Darien, where vast supplies were available. A quantity of Quipo was cut and sent to de Havillands in England. The Quipo test was then carried out at Cowcaddens where a complete fuselage was successfully built up. The plan to build with Quipo was unrealistic, since there were vast stocks of balsa readily available in the country. The Quipo fuselage Mosquito was not put into production.

As an insurance against the failure to supply metal for light bombers Armstrong - Whitworth designed a twin-engined bomber,

The Albemarle, built around the Bristol Hercules aero engine. The design was unusual because it was built around modular form by firms with no aircraft manufacturing experience. Much of the work on this bomber came to Glasgow and the West of Scotland. The centre section was of tubular steel, the steel being supplied by Singers of Clydebank, while the outer wing main spars were built up of spruce members and plywood panelling of laminated compressed wood with tubular steel extensions. Morris supplied A.W. Hawksley all the Albemarle wings, ailerons, tails and vertical tail surfaces for three hundred aircraft. Only about six hundred were made of this tricycle undercarriage aircraft since, by 1943, Britain had a plentiful supply of conventional bombers. Many were used by Transport Command as glider tugs for the invasion of Europe, while a dozen were sent to Russia. The group built Albemarle was not a success. It handled poorly in the air and, when in storage, the wooden wings held vast quantities of rainwater, no drain plugs having been specified for the flaps, ailerons, or elevators. It was notorious for vibration, navigators could not read their charts and its ability to wallow in the air and its lack of trimming ability followed its notoriety.

Early in 1938 the Westland Company had won the contract to supply the Royal Air Force with an army co-operation aircraft, the Lysander. Their chief designer was Arthur Davenport, whose design was very similar in some ways to Sydney Camm's Hurricane with a tubular framed fuselage with wooden stringers, formers and longerons.

In addition to supplying the technical information on wood to Petter at Westland, H. Morris & Co, supplied wood for the Avro Anson, the de Havilland Dominie 1 (Rapide) and for the Boulton Paul, twin engined, heavy multi - cannon, turreted bomber destroyer which was sub - contracted to Heston Aircraft as a flying model. During the war Rolls -Royce sold their share of Phillips & Powis and the company became Miles Aircraft. They went on to produce the

Master III, Monitor, M. 20 and the Martinet. All of these aeroplanes flew with Morris wood. Even the humble Tiger Moth trainer and the Airspeed Oxford were wooden and flew with Morris frames and panelling.

The glider programme was one of the easiest to develop for the Ministry of Aircraft Production. The glider was a single wing all - wood aircraft which did not require an aero-engine for parallel development. The airborne forces requirement for a glider was also very simple and their specification did not require constant modification. There was also the added bonus for the ministry that the army made up its mind easily. Neil Morris supplied the know-how on glider wood to the Director of Scientific Research and the Royal Aircraft Establishment. This in turn led to close collaboration with Slingsby, Airspeed and Harris Lebus on the Airspeed Horsa and later the Hamilcar tank - carrying glider designed by General Aircraft.

The Miles Falcon, single-engined aeroplane, was used to test the wings and tail of Britain's first supersonic designed aircraft, the Miles M.52. The wings and tail of the Falcon were truly razor sharp in design, giving the aircraft the nickname "Gillette Falcon". Morris made all the test models for the M.52 project, but the type was cancelled in February 1946. It was found that the M.52 was unstable at high speed and that the pilot would not be able to escape if there was any trouble. Another factor which affected the design was lack of finance in the post-war world so, reluctantly, the type was cancelled, but not before Britain handed all the M.52 data to the Americans. All the data was incorporated into the Bell X 1, making it the first truly supersonic design. Vickers resurrected the model as a high speed rocket under the direction of Barnes Wallis. In 1946 these wooden rocket bombs were tested near Loch Long at Loch Striven, with Harry Morris and Barnes Wallis being present. Mr Neil was an observer. Later, Wallis perfected the design into a long-range rocket bomb, which was a post - war development. Morris supplied Vickers with ten flying models and, when the completed model flew in 1948, it was a complete success. The missile was named the Vickers Trans Sonic Missile, flying with Morris mahogany wings and tail and its rocket system a copy of the German Schmetterling missile.

H. Morris & Co. supplied plywood and aeroplane wood to English Electric, Brush Coachwork, Fairey Aviation, John Compton, Vickers Armstrong and Scottish Aviation etc. Scottish Aviation had been entrusted with Liberator modifications at Prestwick and the Morris Company supplied all the flooring and wood modifications for the modified Liberator, wood replacing all the redundant metal in the airframe. This included floors, hatches and panels. Probably the most unusual combination of the Second World War was the Albemarle (Morris wings, tails, spars, etc.) towing Horsa gliders (Morris wood) carrying paratroopers with rifles made at Cowcaddens.

The Scottish Aviation target aircraft which was *the radio controlled de Havilland Queen Bee. Morris received a sub contract to build the aircraft at Milton Street. They in turn sub-contracted the work to the Scottish furniture industry since they did not have the production capacity or the room. They were manufactured in West Campbell Street, Glasgow, though the company did manufacture the floats at Milton Street. The fuselage was all wood with the wings and tail of the Tiger Moth. Some of their retired employees said the fuselages were made on concrete moulds like the Mosquito. Scottish Aviation received the prototype which Morris copied. The aircraft was then sent to America and was used for filming, "The Spirit of St Louis," after World War 2 as NC726A. The Americans were so impressed with the concept they went on to develop the drones - a direct line to the Queen Bee .One Scottish Aviation/Morris/West of Scotland Furniture makers, Queen Bee, LF858, has been restored to flight in the UK*

James Lithgow

Lord Swinton

Lord Weir, was a somewhat secretive figure behind the government in the nineteen thirties and key member of the Committee of Imperial Defence a great friend of Harry Morris.

The de Havilland Mosquito

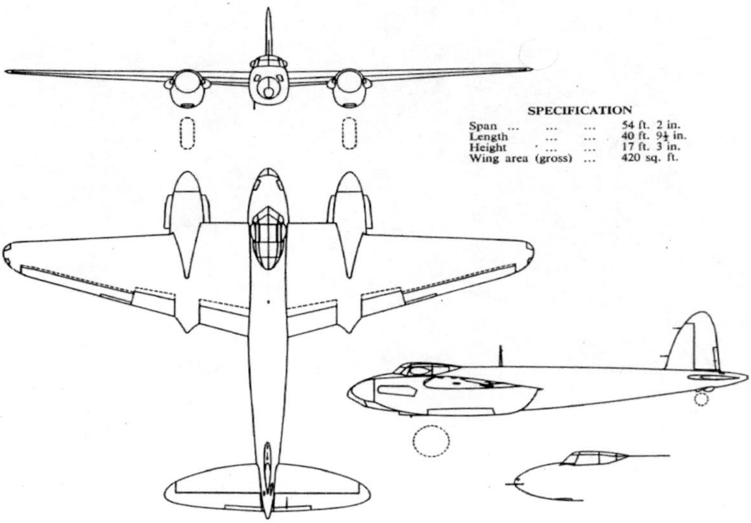
SPECIFICATION
Span ... ... 54 ft. 2 in.
Length ... ... 40 ft. 9½ in.
Height ... ... 17 ft. 3 in.
Wing area (gross) ... 420 sq. ft.

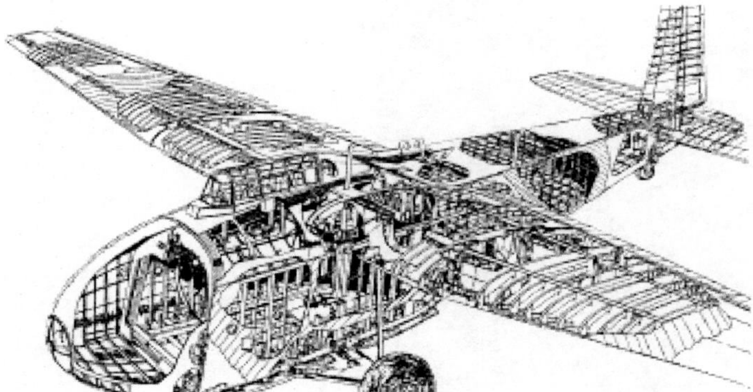

The General Aircraft Hamilcar transport glider of all wood construction, the components were sub contracted throughout the furniture industry and assembled by railway companies.

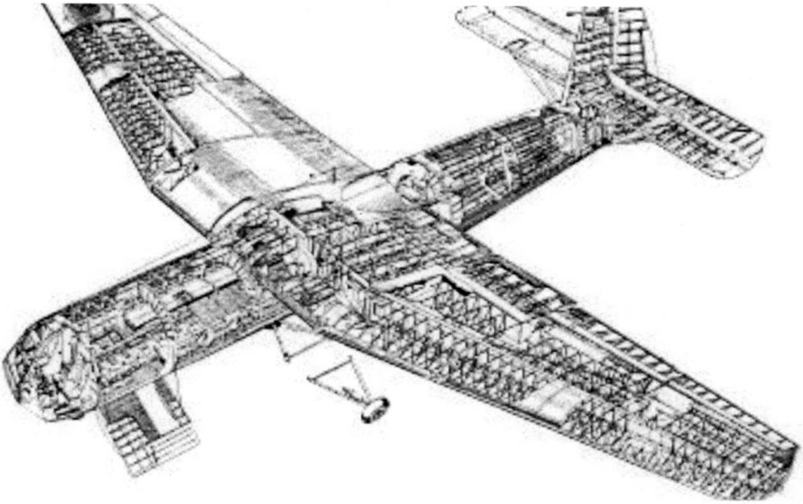

The Airspeed Horsa glider for the assault on Europe. Production was so successful that it exceeded all expectations, overproduction had to be stopped by Churchill.

Making Horsa components at the NBR works Springburn

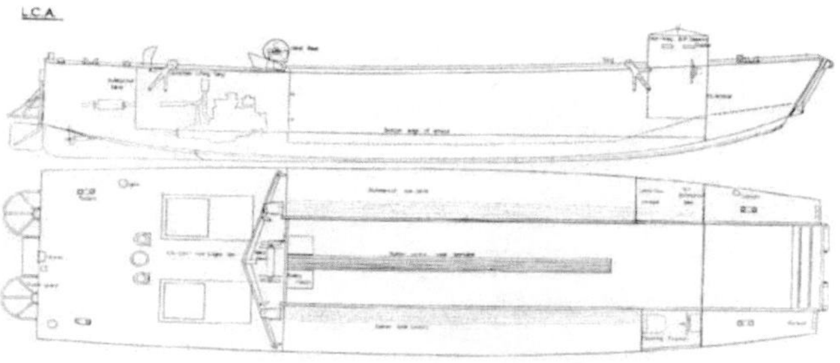

In addition to aircraft, Morris plywood was used to make landing craft

A motor torpedo boat built with Morris wood

A landing craft, June 1944

Converting a Liberator to C-87transport standard and a modified Liberator with single fin

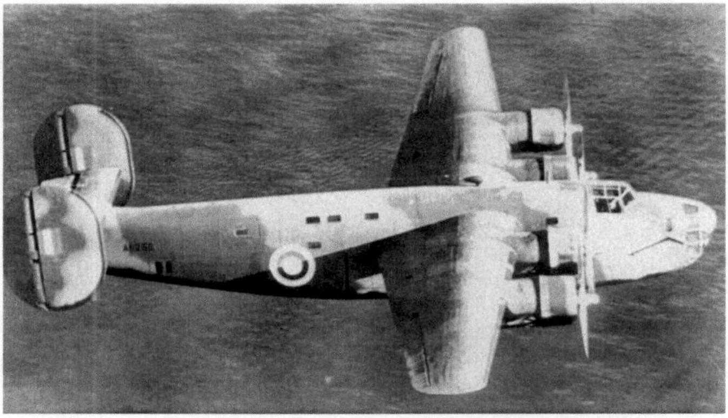

An early Liberator in flight, converted to a transport for returning pilots to America on the Atlantic Ferry run

A Liberator converted for Atlantic patrol with a gun pack and radar at Prestwick

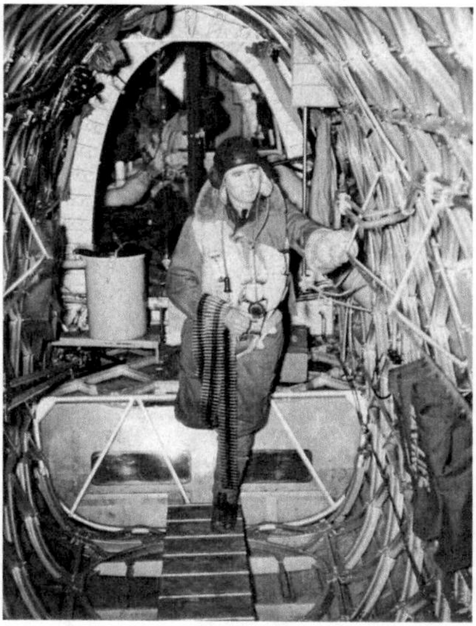

Morris tested the concept of wooden geodetics for the Vickers Armstrong Wellington. This was for the Royal Aircraft Establishment

The interior of a Wellington showing clearly the Geodetics. The fabric was held to the fuselage by plywood panels.

The bonded Spitfire using the Redux process Morris was consulted in its process

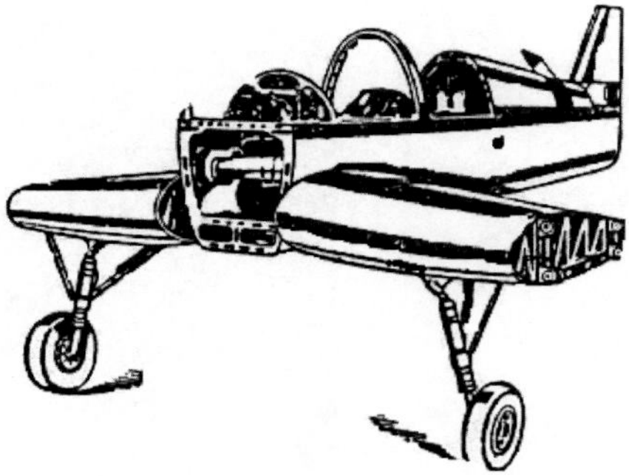

The all-wood Miles Master training aircraft, a masterpiece in wood

The Rolls-Royce Kestrel powered Master in flight

Master in flight with either a British or American radial engine

The Miles Magister all wood two seat trainer

The last Hurricane being assembled in 1944, note the tubular steel structure surrounded by plywood frames and longerons

"The Last of the Many"

A Canadian Hurricane

A Hawker Hurricane in the field.
Morris was a major supplier of aircraft plywood to the manufacturers

The Avro Anson in flight, Morris supplied all its plywood to Avro at Manchester

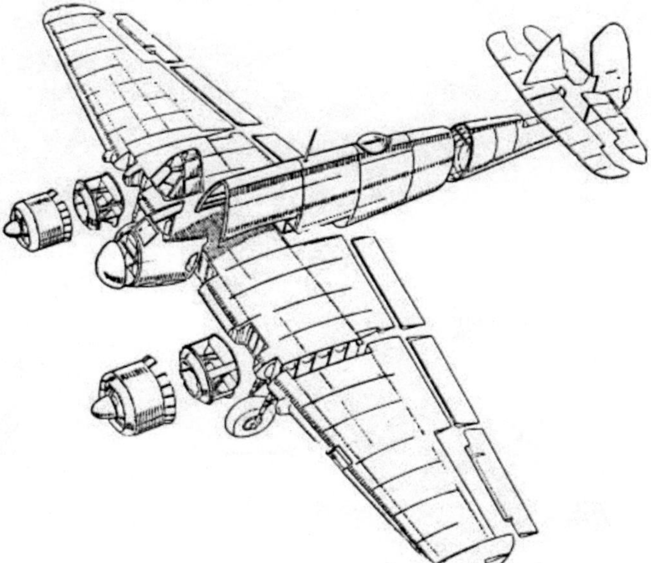

The component breakdown of the Bristol Beaufighter

The Beaufighter as a night and strike fighter

The de Havilland Albatross all-wood airliner

Wilfred Freeman on the left with the Minister of Aircraft Production, Stafford Cripps

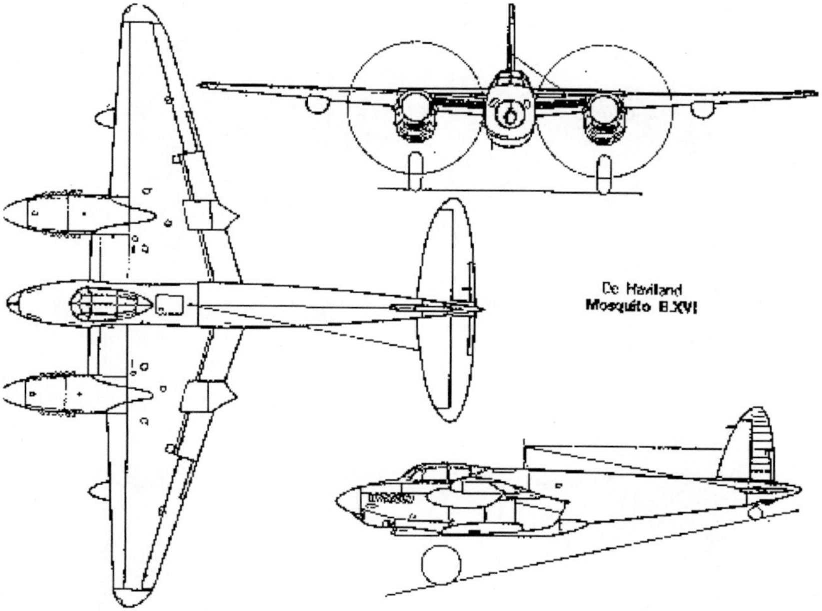

A reconnaissance Mosquito showing the clean lines and the Morris overload fuel tanks

The de Havilland Mosquito fuselage just out of the mould

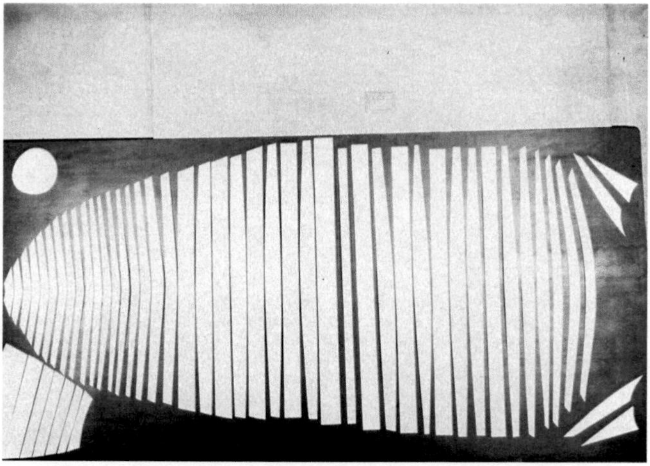

Laminations for the overload fuel tanks.

The Mosquito under construction

Derelict Mosquito, years in open storage, proving the quality of Morris wood. This aircraft was burned in 1963.

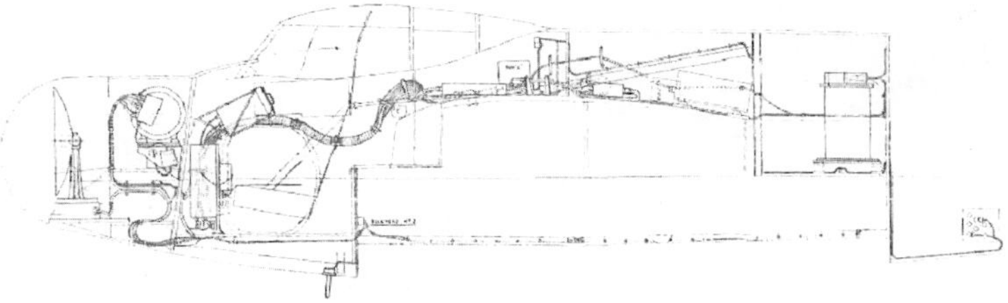

The night fighter Mosquito showing the installation of the radar on the wooden frame, Morris were involved in testing the stress calculations of such installation for the Royal Aircraft Establishment

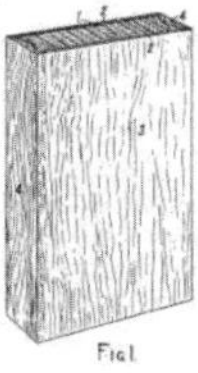

Fig1

The balsa sandwich which Harry Morris patented in 1935. This was the wood used in the Mosquito fuselage

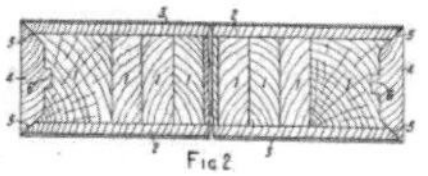

Fig2

An Albemarle wearing invasion markings in 1944 Morris supplied parts for 300 aircraft

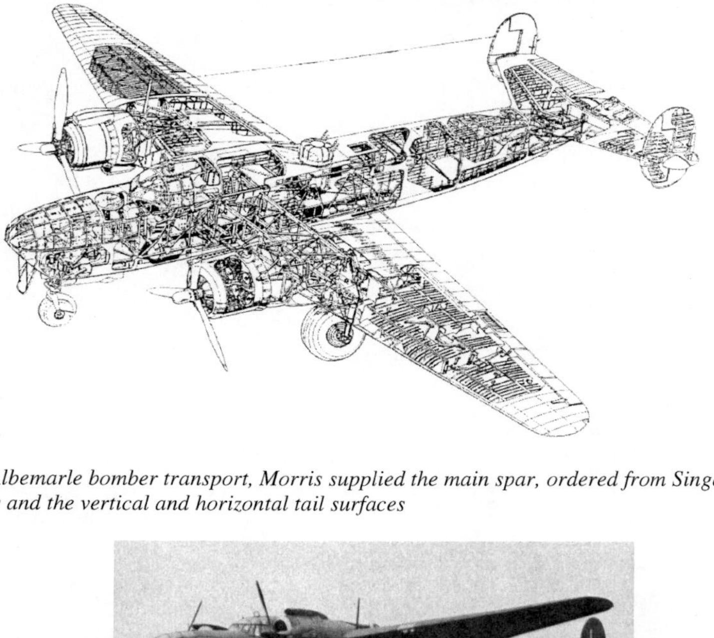

The Albemarle bomber transport, Morris supplied the main spar, ordered from Singer, the wings and the vertical and horizontal tail surfaces

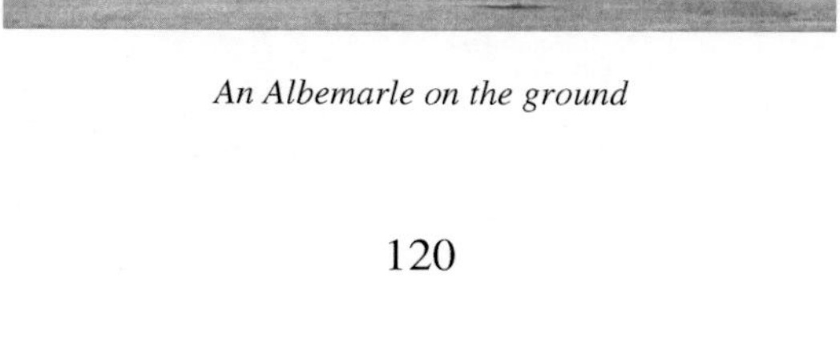

An Albemarle on the ground

A Westland Lysander

The Airspeed Oxford

The Tiger Moth

Hamilcar under Halifax tow and showing a Horsa glider interior

A light tank leaves a Hamilcar

An American Horsa being used to train paratroopers

A wooden mock-up of Britain's first supersonic aircraft, the Miles M52 of 1944

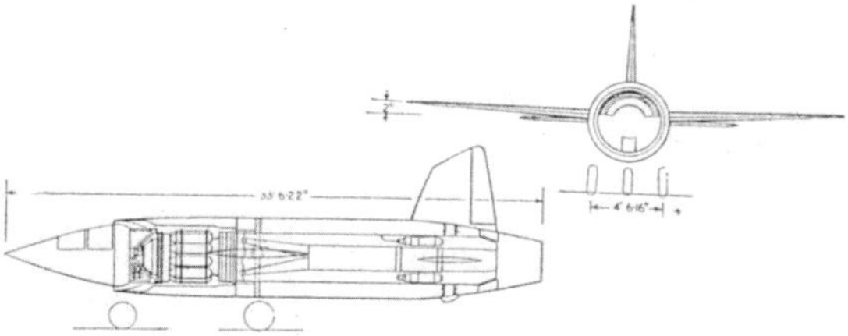

M52 supersonic fighter

The projected M52 in the air

Barnes N.Wallis, designer of a whole range of unusual aircraft and weapons.
He visited Cowcaddens at least twice in 1942

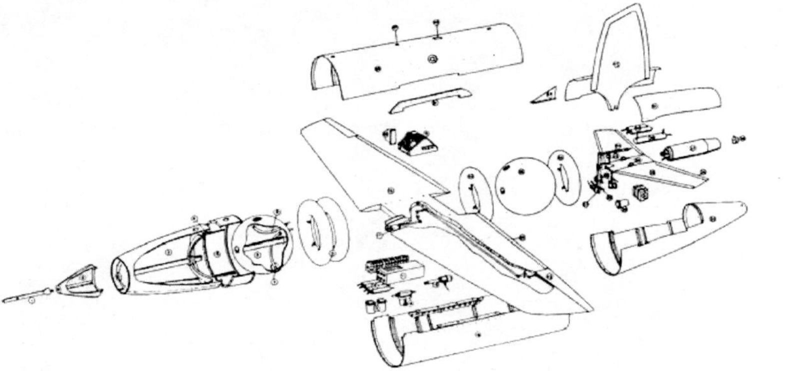

The R.A.E.-Vickers transonic rocket model.

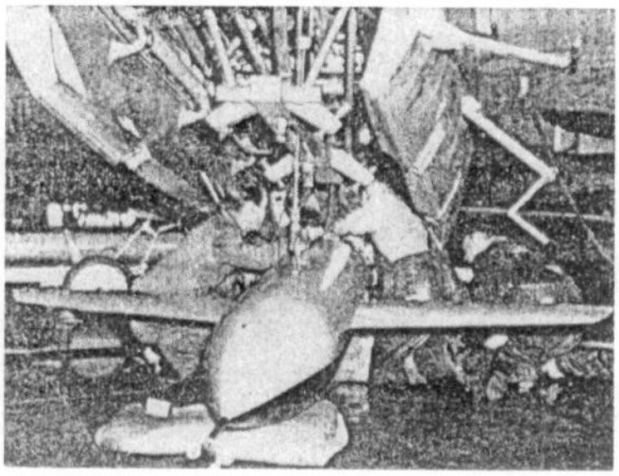

Loading the missile and the missile in flight

A modified Liberator with a radar nose .The success of the Morris conversion meant defeat for the U boats attacking merchant shipping

AEROPLANES

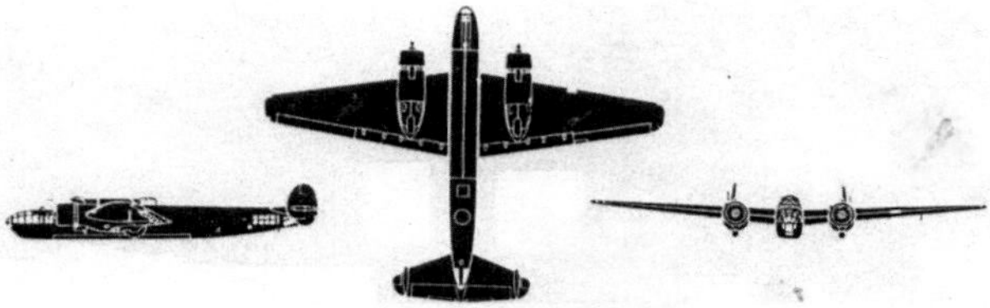

ALBEMARLE

HALIFAX III

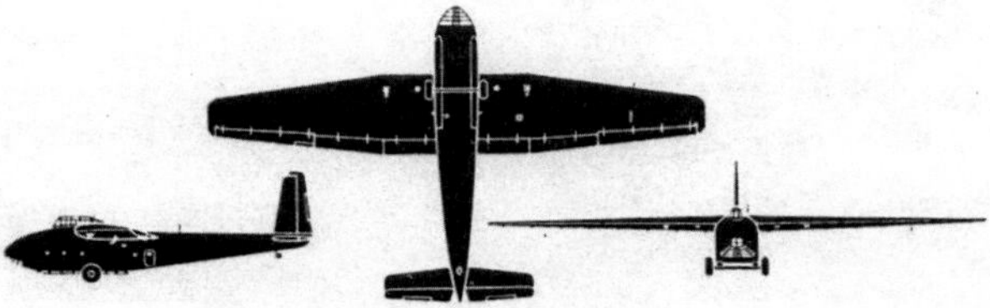

HAMILCAR

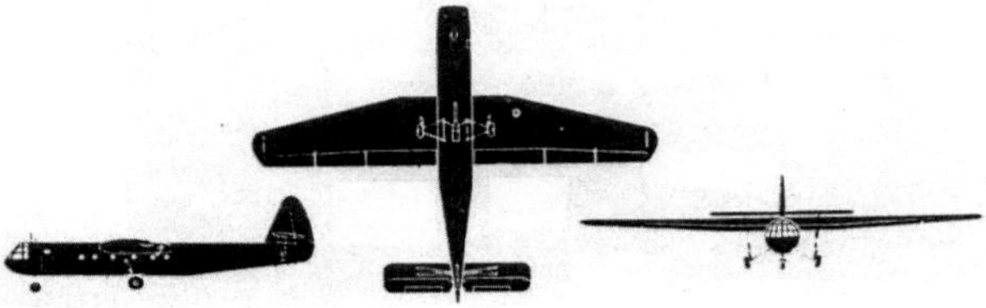

HORSA

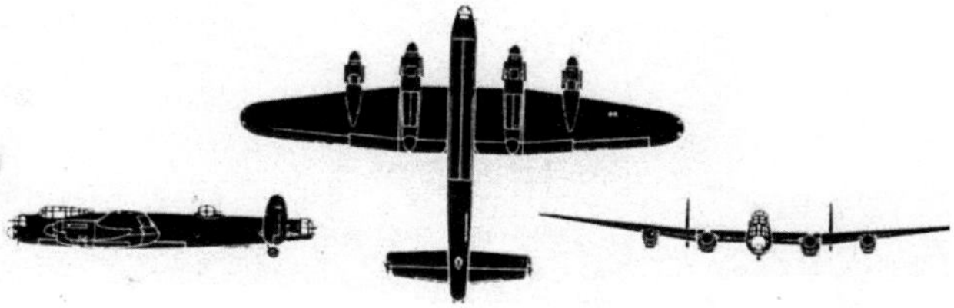

LANCASTER

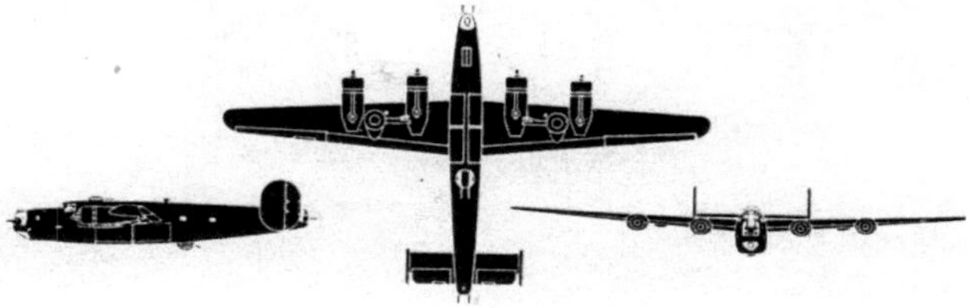

LIBERATOR, B-24

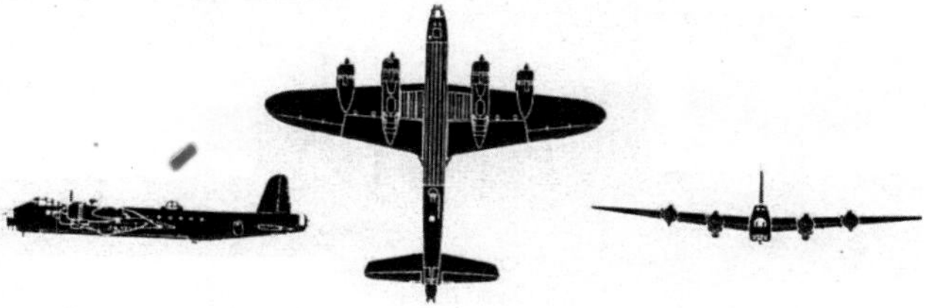

STIRLING

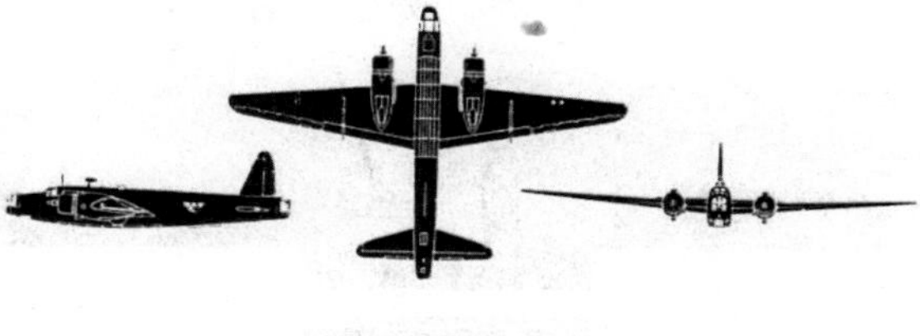

WELLINGTON

Aircraft with Morris wood

HURRICANE

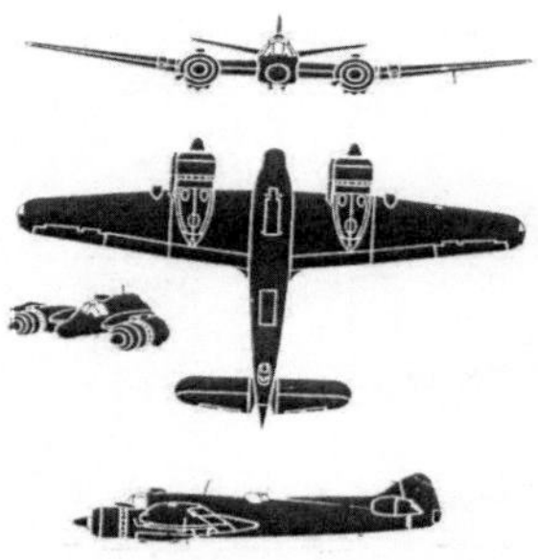

BEAUFIGHTER TF Mk. X

DOMINIE I

LYSANDER

MOSQUITO B Mk. XVI

SEA OTTER Mk. I

TIGER MOTH

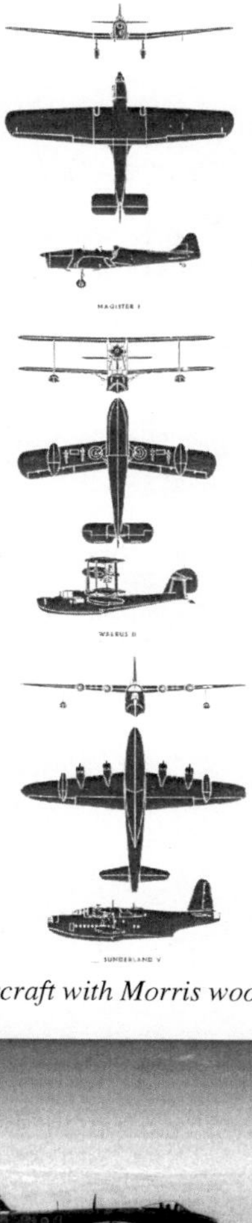

Aircraft with Morris wood

Morris supplied the wooden air intakes for Britain's first operational jet fighter, the Gloster Meteor. The intakes came in two diameters made of mahogany

(iii)

Specific Aircraft Contracts

Company	Date	Ref	Remarks
Gloster Aircraft Company	27 August 1940	A.M.C No B53250/39	Albemarle main spars
C.L.E. (Ringway)	30 July 1941	SB2133/C 20 (a)	8 blades for Dr. Bennett
C.T.O. (Sherbourne)	1 June 1943	SB24944/C20 (a)	For R Malcolm, blades to Hafner drawing R46/55/1452
Cierva	27 May 1944	SB 151216/C20(a)	Blower for W.9
Cierva	26 June 1944	as above	Blower for W.9
Cierva	5 September 1944	as above	W.9 moulded wood tail section.
Cierva	14 September 1944	as above	W.9 blower for Mr. Nisbet
Cierva	9 October 1944	SB 151216/C20(a)	3 sets of W.9 blades
C.T.O. (Beaulieu)	11 January 1945	SB24944/C20 (a)	7 Blades
Cierva	22 August 1945	SB15216/C20(A)	3 rotor blades
R.A.E	22 August 1945	aircraft 4225/DJ223/C20(b)	Model E.6/44
Cierva	26 August 1948	C/AIR/704 CB(b)	3 blades No4 Red No5 white No6 Blue
Cierva	13 April 1949	as above	Verbal order from Mr Shapiro for 3 ply moulded skins
Cierva	25 May 1949	as above	1 Set blades repaired and modified
Cierva	22 September 1949	as above	Blades for W.11

More sawdust on the floor

The woodworking ability of Morris was clearly shown by the amount of one - off projects it was able to undertake during the war, with the barrage balloon being one such example. Originally the balloon was used for gun spotting by units of the Royal Artillery and had evolved into a three finned, captive barrage balloon for use as passive defence. These balloons were placed around strategic targets in Britain and belonged to units of Royal Air Force Balloon Command. This command was under the control of Sir Frederick Pile, who felt that the latest balloons were too labour intensive and difficult to control. In 1940 an order for balloon fins was received at Cowcaddens. The new balloon was very much more streamlined in shape and had three plywood fins mounted at the tail. This was the Mk VI Kite Balloon. It was very much a success and rendered obsolete all the ones that were still in production, including the Mk XIII made at the Kelvin Hall in Glasgow. It needed little maintenance, was easy to make and required little in the way of manpower. Many of them were used in the East Coast convoys and for protection of the "D-Day" Armada in June 1944. After the war they were used as distance markers for the Gloster Meteor, which won the world's speed record for Britain in 1946. Much of the work done with the wood in those balloon tails was done by women war workers and it was up to Charlie Sim and Davie Wright to see that the work was completed to a high Ministry of Aircraft Production standard. When the work was inspected and cleared for aircraft use, the wood was stamped with a seal. The seal was circular and marked, "H.M. & Co" in green.

Alexander McWilliams probably had the most secret job in the factory and he never knew it. Early in 1942 Barnes Wallis arrived at Cowcaddens with C.A. Oakley, the representative of the Ministry of Aircraft Production in the Glasgow district. They were both taken to the office; the doors were shut and the blinds drawn. Subsequently

McWilliams and Peter Blythe were sent for. They were to make a round hollow wooden cylinder, shaped like a cable carrier and inside was to be placed an explosive mine. It was to be strong and light and be able to withstand heavy dropping. Not knowing the full purpose of what they were making McWilliams set to work in a corner of the factory with some of the lady workers and Peter Blyth as his assistant. Barnes Wallis came back and spoke to Blyth and McWilliams with Harry Morris present. Could the wooden cylinder be made slightly bigger, like a sphere, with additional wooden pieces wrapped around the inside? Could it also have a variety of wooden spheres, even a plywood sphere? To prove the design, a tank was installed in the attic of Milton Street. It was the length of the roof and along its length, cameras were installed with special lighting. Further input came from Rosyth and the Mine Warfare Experimental station at Fairlie on the Clyde. This mine was initially code named, "Johnny Walker" and all the paper work was covered with the Johnny Walker codes. It was already in production, with Cowcaddens being heavily involved in its manufacture. In service it was later termed "Highball" and was to be dropped by Mosquitoes on enemy battleships. The mine was installed in the Mosquitoes of 618 Squadron, which flew the type from Prestwick and Turnberry and sometimes Ayr, to Loch Striven, which is opposite Bute, on the Firth of Clyde. A Vickers Wellington was also used on the trial. At the end of the Loch, the Battleship Malaya was heeled over to reveal her hull. As the Mosquito came in, the Highballs were dropped in pairs with all the action being filmed. Harry and Neil Morris were present at some of those trials, sharing the same observation boat. Earlier, an obsolete French battleship, Courbet, was used before Malaya went into trials at Loch Striven. There was no operational use made of Highball in Europe, though there are unsubstantiated reports of 618 attacking a convoy off the coast of Norway. The Mosquitoes and Highballs were shipped to the Pacific from the Clyde, but no use could be found for the type and

Highball was destroyed in Australia. It was up to the Americans to develop the series of spherical bombs, which they called “Speedee,” Britain pulling out of the development work. America was no stranger to bomb work having designed its own spherical mine and completing earlier the offset tail design now ascribed to Barnes Wallis. Baseball was another development, which was fired from the torpedo tubes of Motor Torpedo Boats, but its development came to nothing. Neil Morris accompanied a naval crew to retrieve one of the prototype bombs. As the bomb was recovered he was told he was very fortunate, since the fuse had not fired the charge and they were all lucky to be alive. As a memento of this close encounter with fate he was allowed to keep the naval duffle coat given to him by one of the crew. This was not the only bomb work completed for Vickers during the war. George Hezeltine, of Vickers, designed a whole series of bombs, with extended fuses for overground bursting and to test their aerodynamic qualities Morris of Glasgow manufactured many bomb models. These models were tested at the Royal Technical College wind tunnel and at Farnborough. Other model bombs were made to test fin design and nose design, culminating in the huge "Tallboy" and "Grandslam" super bombs being manufactured in Scotland by Clyde Alloy of the Colville Group and the Steel Company of Scotland. These heavy weapons should be regarded as joint Anglo – American, since they were developed during and after the war by both nations through Lease Lend funding. A much larger bomb was manufactured as a prototype by the North British Railway company at Springburn. It became the 44000lb bomb developed by the Americans.

As a diversion from war work, the company were called into special duties by Timber Control. In February 1940 the old cruiser, H.M.S. Dunedin, stopped the German vessel Hannover and searched for contraband. She was boarded and taken to Blyth. She carried a cargo of plywood, which was taken to Glasgow for inspection. There it was classified and reissued to the industry. Other plywood, from captured enemy vessels, had their cargoes inspected by Morris and cleared for distribution by Timber Control. The German ship Hanover was renamed "Empire Audacity" and converted to an aircraft carrier and while she was at sea, escorting a convoy in the middle of the Atlantic (December 1941) she was torpedoed and sunk, her captain going with her to the bottom.

The sea was again to give another diverse interest to the company. In 1944 an order was received at Cowcaddens from Helensburgh, the home of the Marine Aircraft Experimental Establishment (M.A.E.E.) The M.A.E.E. was once again asking for some unusual help. In addition to small bomb carriers and a wooden catapult of an earlier order, this new contract was to test all the skills of Morris. The war in the Pacific was at its height and the Allies wanted a new fighter to support their island hopping operations. While the new fighter idea was being built Metropolitan Vickers had developed a small jet engine called the F2/4. Later this turbo jet was called the "Beryl". At Saunders - Roe the chief designer, Henry Knowler and his assistant M. Brennan combined the Beryl jet engine and the flying boat into a small twin engine jet fighter. Called the E.6/44, after its specification, it was to be a spectacular performer, but before it could be completed the M.A.E.E. wanted to see if the design was right. Peter Blyth and George Patterson were called to see Mr Neil. He explained to them that the Air Ministry wanted a working two thirds scale model of a small flying boat. The pair, oblivious once again, due to the nature and secrecy of the project, set to work. By the summer of 1945 the model was completed; it was identical to the full scale Saunders Roe design and had small servomotors for the controls. Initially it had a small round tail but when it went into the wind tunnel at Farnborough the model was sent back to have the tail modified to a squarer, taller, shape. From Helensburgh one model was taken to the Denny Tank at Dumbarton for water trials. There it was found that the model was in many ways, superior to the original article's specification and the

Saunders Roe E.6/44 was completed in the light of experimentation of the model. This was the very first test made by Farnborough on a model of an aircraft that was not yet in production, or had not left the drawing board. Morris made three models of the Saunders Roe E.6/44 and pioneered new ground in experimental work on modern aircraft. One model had rockets fitted to it and was flown at Garelochead. Unfortunately the radio controls were faulty and the model crashed. This small fighter was way ahead of anything in its time. It handled well and could be looped and dived at will, with very little or no seawater entering the jet intake. To prove the handling Geoffrey Tyson, the test pilot, flew the little jet at Farnborough, across the airfield, upside down. Even today, it is acknowledged that the Saunders - Roe E.6/44 had an outstanding performance. The only survivor of the three manufactured is in the Imperial War Museum collection and its success in the water and the air speaks volumes for the quality of the experimental working model.

During November 1937 Mr. Muirhead of Scottish Aviation placed a contract for a large supply of aeroplane wood with Harry Morris. At Prestwick this wood was to be used to build aircraft. Later, a radio - controlled aircraft called the "Queen Bee" was ordered by Scottish Aviation. These were made on a sub contract basis at West Campbell Street in Glasgow. In addition to supplying the aeroplane wood, Morris was also asked to supply a large quantity of fixed pitch wooden airscrews for the Queen Bee and Tiger Moth. Originally propellers were made out of laminates of American walnut, but it was found that laminates of mahogany and walnut could be used instead. When the laminates were glued together they were about 0.75" thick and later laminates of birch and oak were used instead. The airscrew was finished by the addition of metal parts, either to the tips of the blade, or the leading edge. Once finished, the blade was varnished and highly polished. The underlying manufacturing process at Cowcaddens was the careful selection and conditioning of the wood.

Even the storage areas were carefully monitored for temperature and humidity and the wood continually inspected.

The fixed pitch airscrew as used in the Tiger Moth and early versions of the Spitfire and Hurricane could not be considered for high performance aircraft. To get the maximum power out of the Merlin engine, the airscrews were made to rotate around the propeller hub to change pitch in relation to airflow. Due to shortages of metal airscrews, the Ministry of Aircraft Production decided to re-introduce the wooden airscrew for high performance aircraft. The other factor to be considered, with the reintroduction of the wooden blade, was accidents. When an aircraft hit the ground the metal airscrew would shock load the engine and in most cases, destroy the power plant. With the wooden airscrew breaking up in a crash, the chances were good that the engine would survive to fly again.

The most important wooden airscrew blades made at Cowcaddens were manufactured from Jablo wood. This was a birch veneer about .75" thick, bonded together with a resinous film called, "Tego Glue Film." The wood board was made with a high density at one end, a low density zone at the root and a transitional zone in between. These boards were heated in a special kiln and with the temperature and pressure gradually increasing, it took about two hours to make one board of Jablo wood. Initially Harry and Neil Morris had to work with simple notes and verbal instruction and though the Ministry of Aircraft Production found the Jablo wood not to their specification, on test the Morris Jablo wood exceeded all expectations. Two other manufacturers involved with the aircraft industry merged to produce propeller blades. These were Rolls Royce and Bristol Aircraft. The company they formed was called Rotol and the blades they designed were manufactured for such diverse types as the Spitfire, Mosquito and Halifax bomber. Morris urgently manufactured 100 prototype blades for de Havilland in 1942 and continued to manufacture Rotol propellers right to the war's end. Mr Orr was placed in charge of propeller production at Cowcaddens.

Just behind the Mosquito fuel tank autoclave production line was the rifle assembly line. At one end of the bench two girls laid out the rifle handholds, while at the other end the girls stacked the wooden buts with Rose McKee doing the weighing and counting. This was the final stage in the production line. Behind the five girls were the plywood packing cases and at the end of the bench was George Yullis who inspected the finished wooden parts. In 1938 Harry Morris had been asked to produce three thousand Lee - Enfield rifle stocks for the War Office as an educational order. (The order was received as a result of the visit of the sub-committee of the Committee of Imperial Defence) The Lee - Enfield was the standard British Army rifle and had been in service, after trials, from about 1907. In its service form it was called the, "Short Rifle, Magazine, Lee - Enfield Mk III" and it was very good: in fact it was very good indeed being mass - produced by the thousand during the Great War. As well as aeroplane wood Mr Neil was in charge of the rifle programme. He had many meetings with the whole work force and pointed out the necessity for accuracy and he was always at the end of the telephone in case of trouble. Mr. Neil's interest went so far as to specify the colour of the paint on the packing cases: it was red!

Initially these stocks were made up of Italian red walnut and were in two pieces. The first piece of the hand guard was held onto the barrel by a steel band and the nose cap. The butt was forced and compressed into the rifle body and was secured by a stock bolt. Inside the butt was the pull through and oil bottle for rifle cleaning. These wooden rifle parts were cut out on circular saws and finished with sandpaper and soaked, with this duty being completed by McCafferty and Tait. Before manufacturing took place the wood was carefully seasoned and stored before any further operations took place on it. Such was the demand for rifle wood that most of the material used was dry kilned. The groove for the rifle barrel was machined to fit and a hole cut for the magazine, while the butt was first cut roughly on a band saw and then machined to fit the moving parts and the butt plate. The finished product was polished on the outside. The success of the service rifle meant that it could be developed into other versions but the main problem was mass – production and standardisation, every working part in every rifle had to be interchangeable with any other mark of service rifle. The new rifle was termed the "Rifle NoIV Mk l" and it was made in exactly the same way as the No.III rifle. It was descended from a modified Mark III called the "No I Mk IV" which was in prototype form in 1931. For this new rifle, Morris was in sub - contract with Slazenger, Albion Motors and the Singer Company of Clydebank. Singer was no stranger to rifle production for, under the authority of Fred Lobnitz, they had produced thousands of rifle parts during the Great War. It would appear that the No IV rifle was produced in exactly the same way as the No Ill, with the finished parts being assembled at the Royal Ordnance Factory at Fazakerley outside Liverpool. There were twenty seven separate cutting operations on the first part of the hand-guard which was in two parts, from rough cut to profiling and on the butt there were seven operations. The works Number 47 was given by Morris to rifle manufacturing, with production rising from 3 - 400% in a matter of weeks, which meant that by February 1943 they had manufactured fifty thousand hand-guards and rifle butts for the War Office. Due to the war, the factory went into a twenty one hour day and worked for five and a half days a week. There was no forty hour week.

Inside a case at Nottingham there is a standard Mk IV rifle made with plastic impregnated furniture. It is a Morris Rifle with Mr Neil being responsible for its development. Using practices developed for aircraft woods Mr Neil made a set of impregnated rifle parts on an autoclave for Fazakerley. Again it was a success, but because of production problems, the design was never used. On re - inspection by

Harry Morris, he confirmed the decision of Neil Morris, but he also added that the new plastic impregnated rifle furniture was too heavy for military use. With the outbreak of the Korean War in 1950 the Ministry of Supply approached the company with a view to putting the No IV back into production. In 1951 the rifle making machinery was taken out of mothballs and serviced for production. The rifle wood was to be produced, not at Milton Street, but at the recently purchased Campsie site as a shadow factory. This order was to prepare the company for production of the new automatic rifle, the EM2, but with the ending of the Korean War and another automatic rifle being tested by N.A.T.O., new production of the No IV was cancelled.

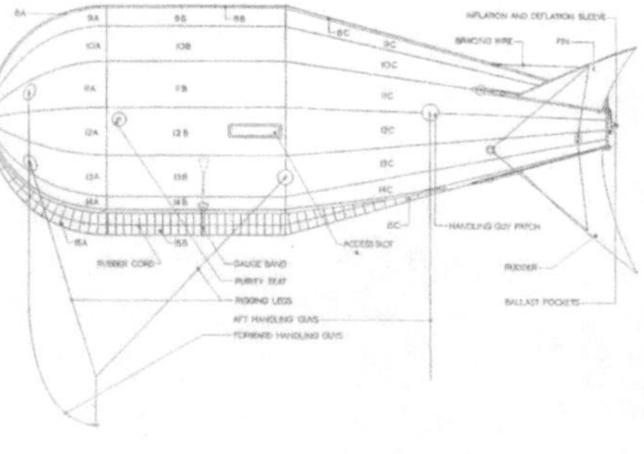

A Mk VI kite balloon and afloat on an East Coast convoy

Kite balloons protect the invasion fleet in June 1944

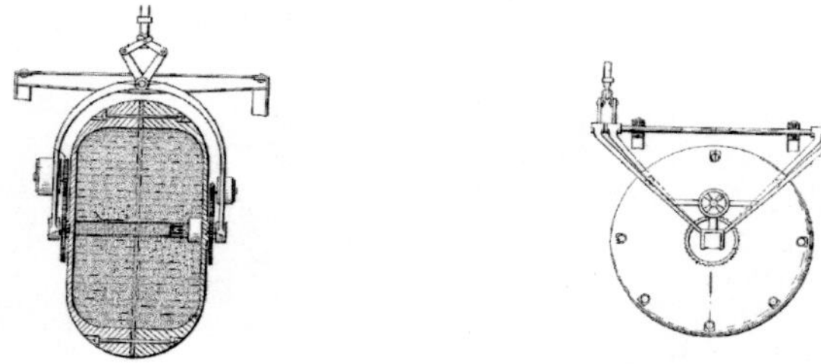

The American bouncing bomb of 1942-

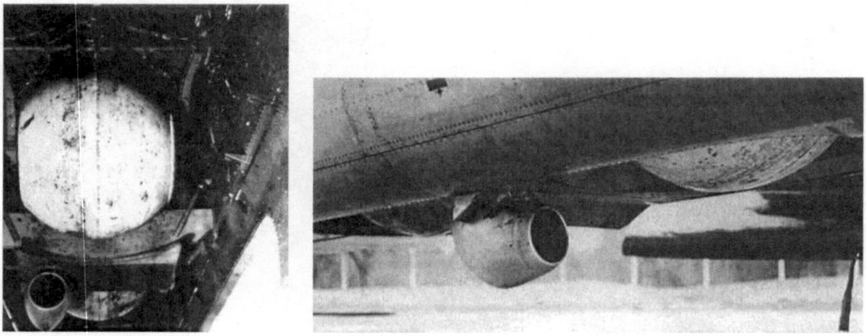

This is the American copy of Highball named "Speedee" The Americans visited Glasgow during the development of these weapons. Initially Morris was involved with the larger version called "Upkeep" of the dam's raid.

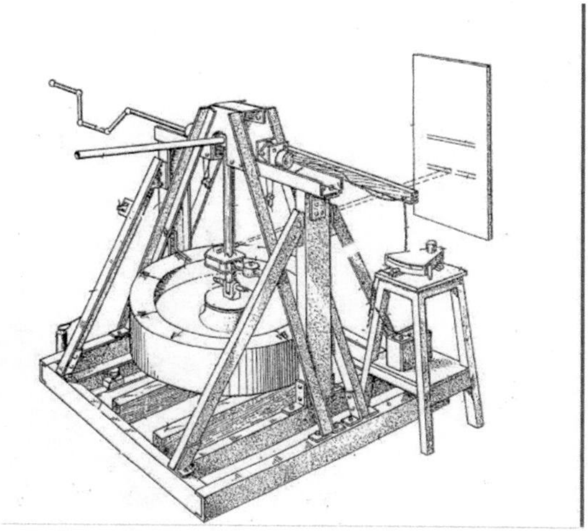

Wallis balance device, employees stated it was for use with Upkeep/Highball device

Upkeep with the wooden casing. This broke on impact with the water and the mine was dropped with just the mine casing, during Operation Chastise, May 1943. Noteworthy in the background is a Mosquito. On visiting the factory with Wallis, Guy Gibson is reputed to have disappeared to look for sweets

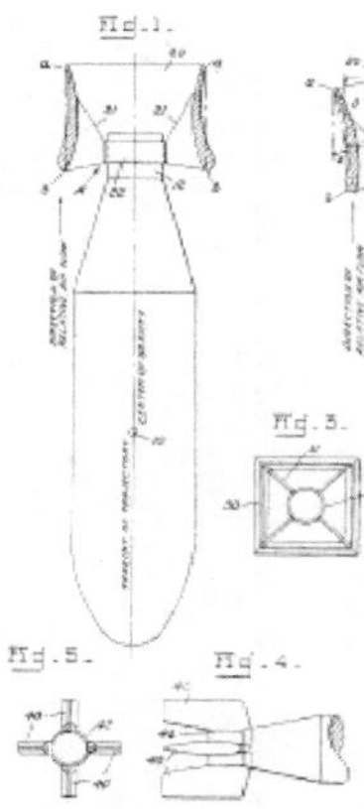

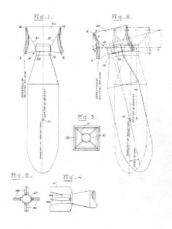

The American off set tail

A together with other patterns in wood

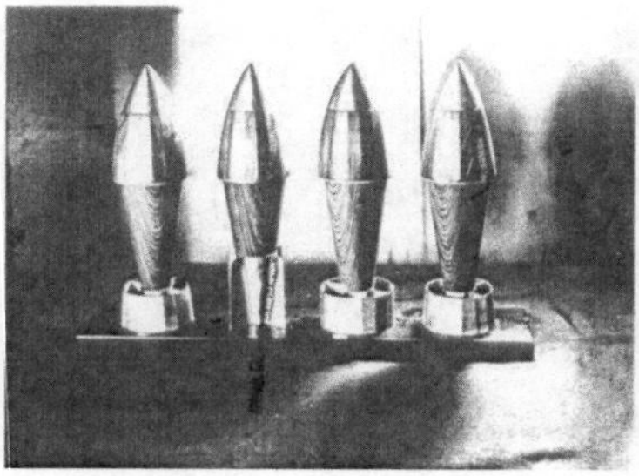

Wooden bomb patterns

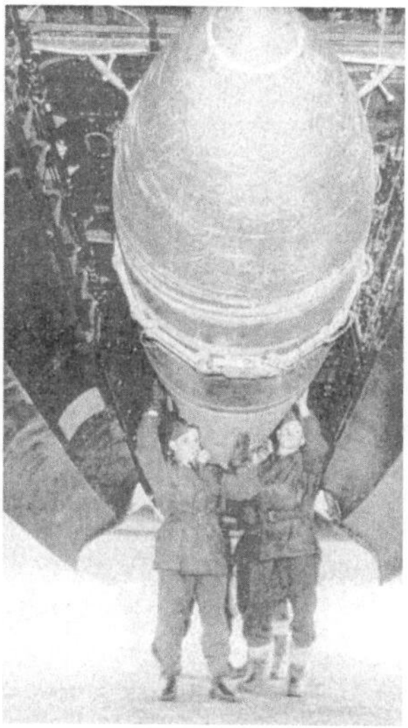

Morris was involved with Clyde Alloy, North British Loco Works and the Steel Company of Scotland in the manufacture of giant aerial bombs. They were directed by the Ministry of Aircraft Production, Ministry of Supply etc.

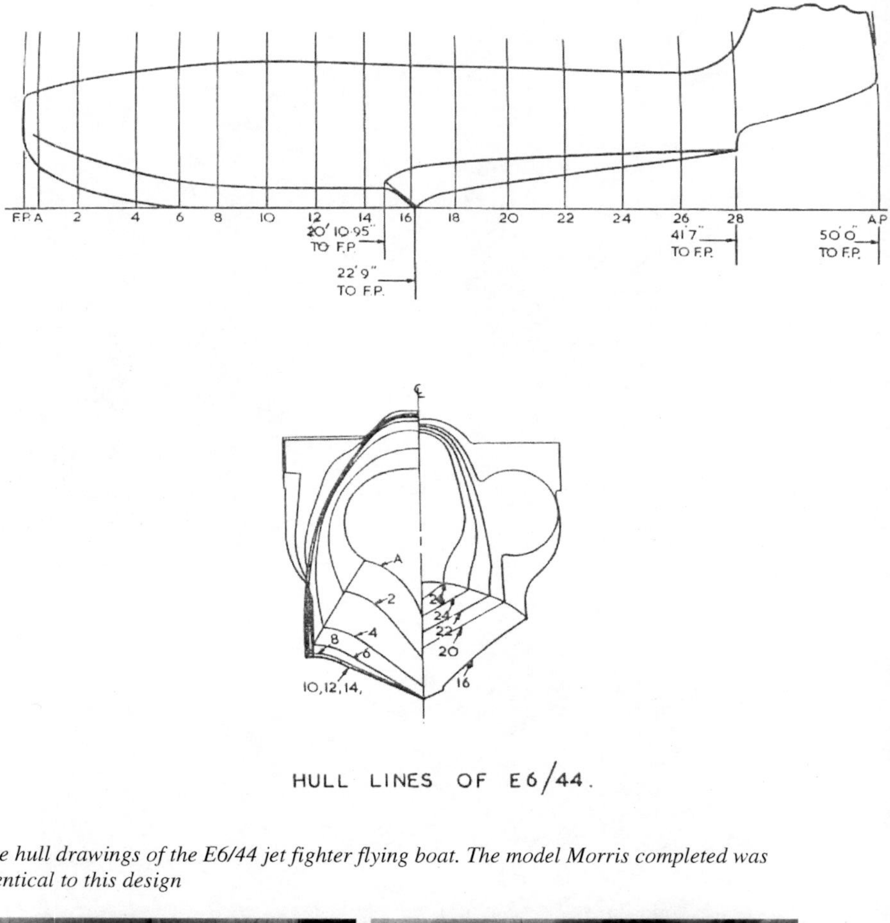

HULL LINES OF E6/44.

The hull drawings of the E6/44 jet fighter flying boat. The model Morris completed was identical to this design

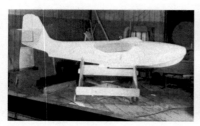

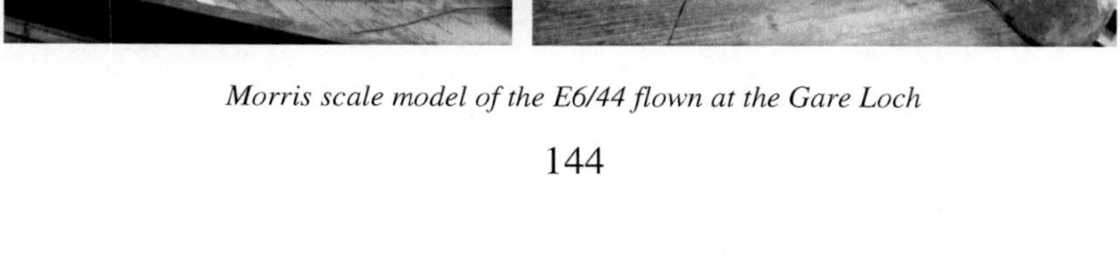

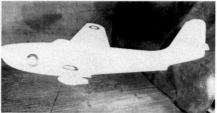

Morris scale model of the E6/44 flown at the Gare Loch

The E6/44 became the Saunders Roe SRA1 fighter

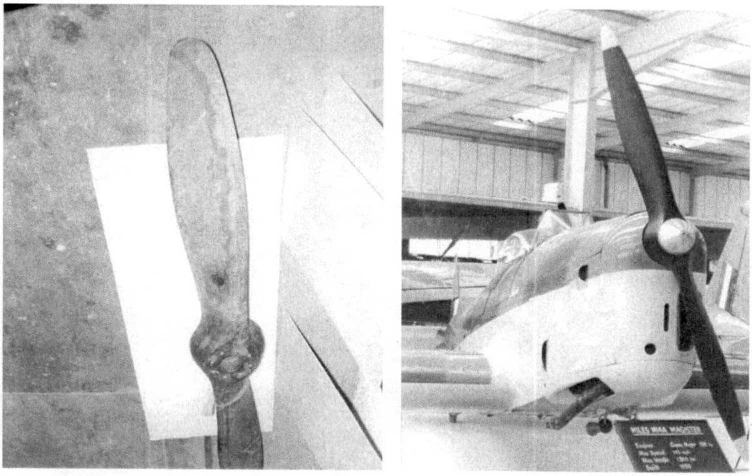

Morris provided propellers for the Miles Magister and de Havilland Tiger Moth. One propeller was found during the move to Drakemire Drive

H. MORRIS & COMPANY LIMITED

BLADES FOR VARIABLE PITCH PROPELLERS

CONSTANT & VARIABLE DENSITY LAMINATED COMPRESSED WOOD

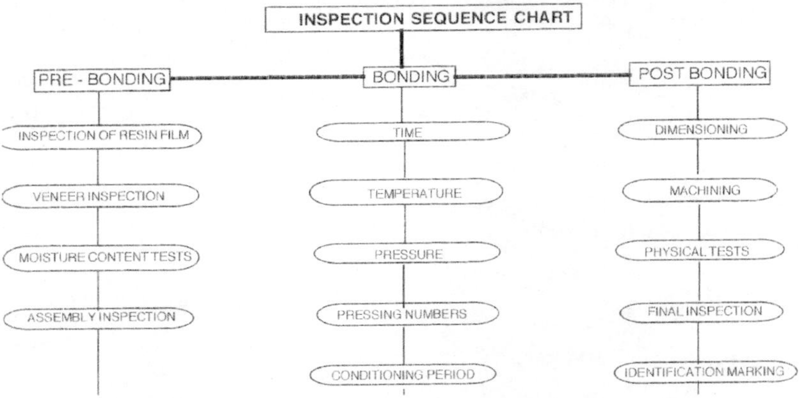

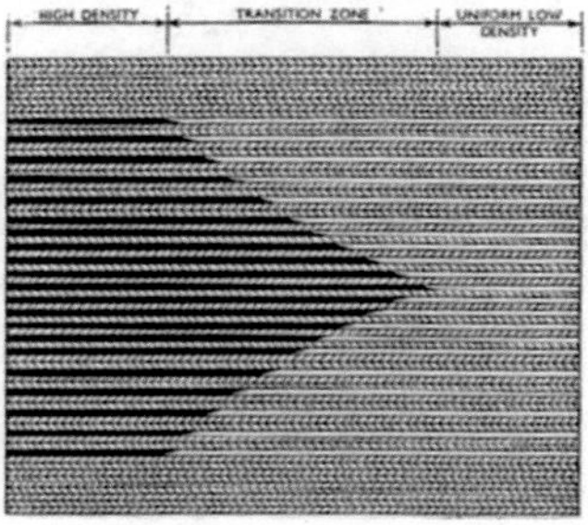

Manufacturing a Propeller and a profile of Jablo Propeller board

The company manufactured blades for the Spitfire and the Merlin powered Halifax bomber

The Lee Enfield rifle of world War One and in World War Two

EM/SG. COPY. 27th January, 1943.

Victoria 3811 Ext.

MINISTRY OF SUPPLY,
(CONTRACTS DEPARTMENT)
GREAT WESTMINSTER HOUSE,
HORSEFERRY ROAD,
WESTMINSTER, S.W.1.

In reply please quote:-
294/R/6175 (Con.10A)

22nd Jan. 1943.

Gentlemen,

Rifle Furniture No.4 294/R/6175 Con.10A.

With reference to your letter EM/EM of 20th ultimo, I am directed to inform you that it is confirmed that the prices for processing and assembling the sets of furniture being supplied under the above mentioned contract are as follows:-

(1) For the first 50,000 sets ordered in the Department's letter of 21st January, 1942, 12/6d per set.

(2.)For the two subsequent quantities of 50,000 each ordered in the Department's letters of 2nd June, 1942 and 16th September, 1942, 10/- per set.

These prices are net, all cash and trade discounts being allowed for, including packing and loading but exclusive of carriage.

Where production is from Home Grown Timber the price covers processing from blanks and the question of the cost of converting into blanks such of this timber as may have been issued in planks is under considera ion as is also the question of kiln drying European timber purchased by you and of kiln drying home grown timber from the green state.

The price of the timber purchased by you for use on this contract, provisionally fixed at £5200.0.0. will be settled later.

/The

Messrs. E. Morris & Company Ltd.,
Milton Street,
Glasgow, C.4.

-2-

Deliveried are to be made in accordance with S.O.44. Transport Arrangements and for the purposes of this clause the Area Transportation Officer's address is

116, Hope Street,
Glasgow, C.2.

Telephone Number:- Central 4104 - 6.

I am to add that for invoicing purposes the figure of 15s. 0d. per set may be apportioned amongst the various components as follows:-

Guards hand front	1s.	6d.	each.
rear	1s.	6d.	"
Stock Butt normal	4s.	6d.	"
Fore end	7s.	6d.	"

An acknowledgment of this letter is requested.

I am, Gentlemen,
Your obedient Servant,

(Sgd) C.E. BAKER.

For Assistant Director of Contracts.

opies given to:-

1 to Mr. Morris.
1 to Mr. Neil.
1 to Miss Strachan.
1 for File.

Original in safe

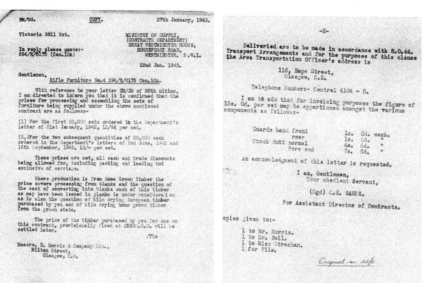

Rifle contracts

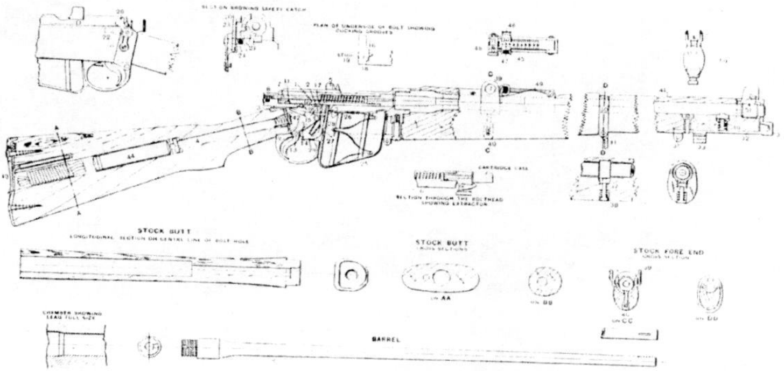
STOCK BUTT
STOCK BUTT
STOCK FORE END
BARREL

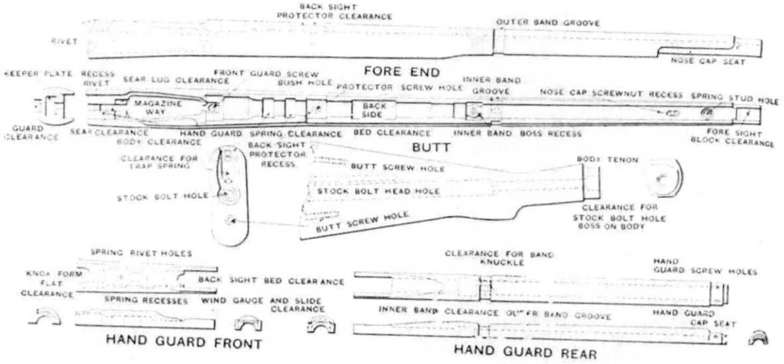
BACK SIGHT
PROTECTOR CLEARANCE
OUTER BAND GROOVE
RIVET
NOSE CAP SEAT
FORE END
KEEPER PLATE RECESS
RIVET
SEAR LUG CLEARANCE
FRONT GUARD SCREW
BUSH HOLE
PROTECTOR SCREW HOLE
INNER BAND
GROOVE
NOSE CAP SCREWNUT RECESS
SPRING STUD HOLE
MAGAZINE
WAY
BACK
SIDE
GUARD
CLEARANCE
SEAR CLEARANCE
BODY CLEARANCE
HAND GUARD SPRING CLEARANCE
BED CLEARANCE
INNER BAND BOSS RECESS
FORE SIGHT
BLOCK CLEARANCE
BUTT
BACK SIGHT
PROTECTOR
RECESS
CLEARANCE FOR
TRAP SPRING
BUTT SCREW HOLE
BODY TENON
STOCK BOLT HOLE
STOCK BOLT HEAD HOLE
BUTT SCREW HOLE
CLEARANCE FOR
STOCK BOLT HOLE
BOSS ON BODY
SPRING RIVET HOLES
KNOX FORM
FLAT
CLEARANCE
BACK SIGHT BED CLEARANCE
SPRING RECESSES
WIND GAUGE AND SLIDE
CLEARANCE
HAND GUARD FRONT
CLEARANCE FOR BAND
KNUCKLE
HAND
GUARD SCREW HOLES
INNER BAND CLEARANCE
HAND GUARD
CAP SEAT
HAND GUARD REAR

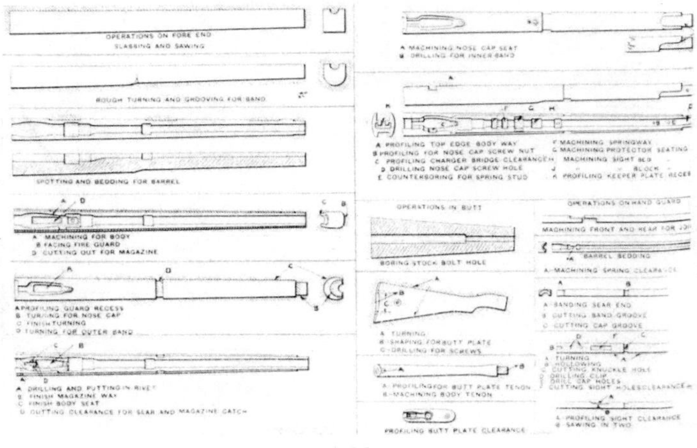
OPERATIONS ON FORE END
ROUGH TURNING AND GROOVING FOR BAND
SPOTTING AND BEDDING FOR BARREL
A MACHINING FOR BODY
B FACING FIRE GUARD
D CUTTING OUT FOR MAGAZINE
A PROFILING GUARD RECESS
B TURNING FOR NOSE CAP
C FINISH TURNING
D TURNING FOR OUTER BAND
A DRILLING AND PUTTING IN RIVET
B FINISH MAGAZINE WAY
C FINISH BODY SEAT
D CUTTING CLEARANCE FOR SEAR AND MAGAZINE CATCH
A MACHINING NOSE CAP SEAT
B DRILLING FOR INNER BAND
A PROFILING TOP EDGE BODY WAY
B PROFILING FOR NOSE CAP SCREW NUT
C PROFILING CHARGER BRIDGE CLEARANCE
D DRILLING NOSE CAP SCREW HOLE
E COUNTERBORING FOR SPRING STUD
F MACHINING SPRINGWAY
G MACHINING PROTECTOR SEATING
H MACHINING SIGHT BED
K PROFILING KEEPER PLATE RECESS
OPERATIONS IN BUTT
BORING STOCK BOLT HOLE
A TURNING
B SHAPING FOR BUTT PLATE
C DRILLING FOR SCREWS
A PROFILING FOR BUTT PLATE TENON
B MACHINING BODY TENON
PROFILING BUTT PLATE CLEARANCE
OPERATIONS ON HAND GUARD
BARREL BEDDING
A BANDING SEAR END
B CUTTING BAND GROOVE
C CUTTING CAP GROOVE
A TURNING
B HOLLOWING
A PROFILING SIGHT CLEARANCE
B SAWING IN TWO

Projected 1950, the replacement for the British bolt action rifle the EM 2

The successor to the EM 2 and the Lee Enfield rifle was the superb Belgian 7.62 SLR, This meant cancellation of the rifle order and of transferring stored rifle machinery to Campsie Glen.

Aquitania on war time duty.

Queen Elizabeth in her wartime colours. She was christened "The Monster" by the Ministry of War Transport

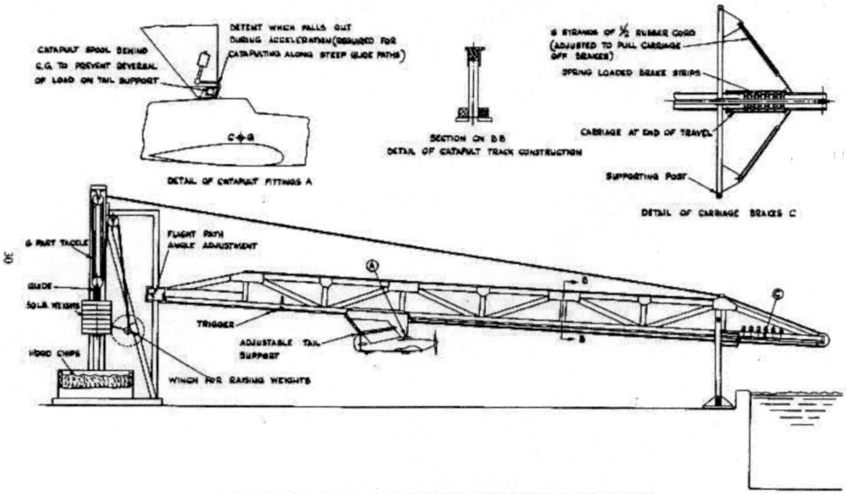

FIG. 2. Details of catapult. Royal Aircraft Establishment Free Launching Tank.

The catapult for the Royal Aircraft Establishment used for ditching trials

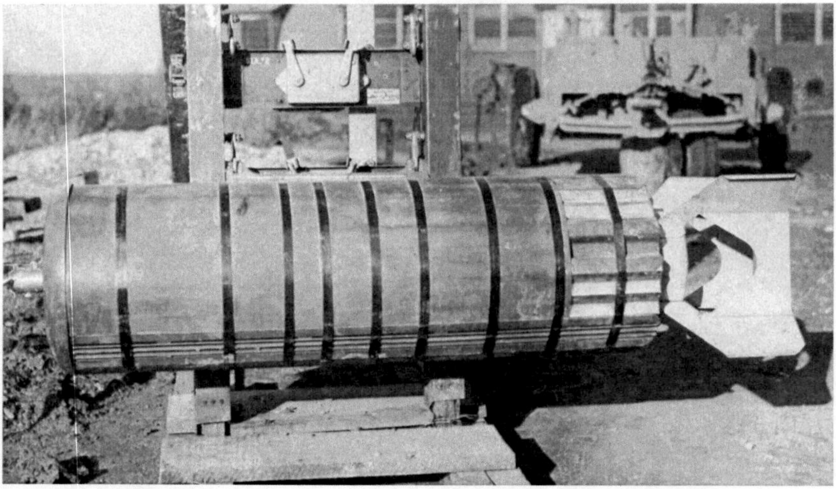

Morris provided the cases for incendiary clusters and early cluster bombs

Peace

Peace came to a bankrupt and exhausted Britain, a Britain totally ruined by the ravages of war and having to exist on the Lend Lease arrangements from the United States. It was a time of austerity and rationing and everything was for export. In November 1945 Harry Morris went to America on one of the first civilian crossings of the Atlantic from Southampton to New York in the Queen Elizabeth. He was to follow exactly the route taken by his son Neil over six years before, even staying in the same hotels. The Americans were astonished at his vitality for his age; he always kept appointments and was always on time. As he booked in and booked out there was always a telegram from Glasgow, explaining the situation at the factory. On one occasion he escaped his itinerary and the representative from United Steel had to go down the taxi rank looking for him. He delighted in being called, "Pop." Harry Morris had gone to America to order fresh machinery to take over from the run-down machines of World War Two. But when he returned in December 1945 he was not a well man. His health started to deteriorate and he was admitted into Hairmyres Hospital. Ever anxious to get on with the job Harry was a difficult patient and tried to leave the hospital, but his health was bad and getting worse. He received letters from the firms that he had supplied for nearly 40 years expressing their concern. Harry Morris did not live to see 1946 into 1947, dying in November of that year, mourned by his family and his friends. The firm passed into the hands of Neil Morris.

Work came in to refurbish the Queen Elizabeth and the Queen Mary, together with the Aquitania, once again working with Mr B.P. Camp. The furniture for these vessels had been stored in the United States, South Africa and Australia and had been returned to Britain. Many of the workers who were sent down to Southampton were women and Messrs Thorneycroft accommodated them in a former

U.S. Army camp. It took sixteen weeks for the refurbishment of Queen Mary and the job was completed on 1 October 1946. The women cleaned and varnished the returned furniture for ship use and in addition new pieces were supplied from Morris stock. Dressing tables, beds, tallboys, nest tables, writing tables, wardrobe doors, and new cabin doors were all supplied from stock. Though these contracts were lucrative and were a ready change from government wartime contracts, there were serious restrictions on the furniture industry in those times. The Board of Trade started to exercise its control over the furniture and clothing industry when it introduced the term "Utility" for a range of civilian goods in 1941. Due to Timber Control, the Board had total control over the furniture industry. In December of that year they issued a list of furniture which could be made and anything that was not on the list was prohibited from being manufactured. The list was enforced by an order in July 1942. The order included only eighteen articles, all essentials, such as wardrobes, dressing chests, pedestal dressing tables, bedsteads, kitchen cabinets and chairs. The timber content was clearly set out for all to see.

In 1943 they produced a catalogue of good designs for "Utility Furniture." It was divided into five sections: living room, bedroom, kitchen, nursery furniture and miscellaneous (which included items like bookshelves and a bed-settee). Each piece used minimal materials and was made of strong and serviceable oak or mahogany, with mortised and pegged joints. Veneered hardboard was used for panelling, since plywood was unavailable. It was required for wartime aircraft production, and most cabinet furniture characteristically stood on plinths, rather than legs. Handles and knobs were of wood, because most metals and plastics were needed for the war effort, although, perhaps surprisingly, metal screws were specified in Utility construction, which added greatly to the strength of the finished furniture.

There was no choice in the matter for the British public. Those who could secure permits had to buy the government furniture. If you had no ration points you did without, or bought second hand at exorbitant prices. Bombed out civilians were permitted furniture as were newlyweds and in some quarters Utility was welcomed because it put a stop to shoddy manufacturing and the under - the - counter black market. It was a seller's market, not a buyer's market. H. Morris & Co were involved in the scheme and at one time manufactured 200000 Utility chairs for the British market, also producing wardrobes and tables to the Board of Trade specification. Before starting production Timber Control authorised the issue of the wood from the wood supplier. The major wood suppliers were Messrs Brownlee and Robinson Dunn at Temple, beside the Forth and Clyde Canal; wood was even stored near Helensburgh, at Rhu and supplied from there to Milton Street. Utility was replaced by a kite mark; the mark meant that the furniture was made to a British Standard Specification (BSS). The kite mark was an indication of excellence, but to stamp your furniture with the kite mark you had to register. The industry was not happy with the scheme and many companies refused to register, H. Morris & Co was one of them. The scheme was a failure and ended with the de regulation of the industry, well into the nineteen fifties and the loss of timber control. Neil Morris extended his skills into interior design, working in houses at Giffnock and the luxury flats of Kelvin Court at Kelvindale in the west end of Glasgow. Neil Morris had the answer to interior design, "Build round any fitment that cannot be moved and do it in such a way that it is practically moveable and suits your way of life." Discerning customers always demanded Morris furniture for the completed house or flat.

The major success for Neil Morris was actually selling Utility furniture to America in 1947. It had taken him two years to secure the American order, Neil felt that, if the order was taken in January and supplied in July, it had become prehistoric and would need updated. He believed in supplying the customers with what they wanted, now.

The Board of Trade took a census of Production in the early post war years. They found that there were nearly 3000 firms involved in the manufacture of furniture, chairs and upholstery, which employed less than ten employees. There were approximately 1000 firms employing more; of these, 485 were mainly cabinetmakers, 191 in upholstery and the rest were mixed firms and amongst the 1000 firms 37 employed more than 300 and only two employed more than 750. Even with the introduction of machinery many of the firms still resembled the origins of its craft industry, hand tools being very much in use. They pointed out a weakness in the structure of most of the firms; many of the functions of management were in the hands of the owners, design, costing, welfare etc. H. Morris & Co had evolved into such a company, with Harry Morris in total control. There was also the conflict of the demands of post-war government and rationing with the demands of the British public. All this coincided with price control and competition from second hand pre-war furniture, with a flourishing black market. The small firm, employing less than ten men was not as subject to scrutiny as the big firm in the class of H. Morris & Co. These firms made small one off orders then quietly disappeared.

Neil Morris wanted to diversify from the large-scale contracts for ships and hotels and move into the household market and office furniture, inspired by the American designer Raymond Loewe. The company diversified into making 100 feet extension ladders for London County Council Fire Brigade. Much of the woodworking machinery had been imported from America, but Neil Morris himself had designed most of it. Kiln drying was a feature of the furniture industry of those times and Neil had designed his own kilns to his own specification with the help of the Forest Products Research Laboratory at Princes Risborough. The greatest problem for the firm between 1944 and 1949 was that of an inability to expand, because there was no land in the vicinity. Neil thought of moving to one of the new industrial estates, such as Hillington. Rowan & Boden had made such a move, but Neil Morris felt that it would be difficult to obtain adequate labour. The company stayed at Milton Street. Though Neil Morris wanted away from ship work the reputation of the company meant that these contracts were to continue for some time. Refurbishment of the Britannic and the building of the Caronia came in. This was followed by work on the Franconia, more doors and beds being ordered for this ship. It took 29.7 hours to manufacture a door and ships such as the Empress of Scotland had to have a separate drawing for each door. The Empress of Scotland was refurbished by Cammel Laird at Birkenhead.

The Royal Naval Reserve former clipper ship, Carrick, was also panelled from Milton Street. She was moored in Glasgow Harbour as a reserve officers' club for many years. As the City of Adelaide, she had sailed on the wool run to Australia in the nineteenth century carrying immigrants and trade goods from Britain. Bar stools were also supplied in addition to the panelling and craftwork. For Canadian Pacific two vessels were refurbished. They were the two surviving Duchess vessels, the Duchess of Richmond and the Duchess of Bedford. It took Morris forty weeks to refurbish the Richmond and thirty-three weeks to do the Bedford. Then an order came for another Cunarder, Scythia which was being built for the Southampton - Le Havre - Quebec service. The first class cinema was designed in the Italian Renaissance style, with seating for 265. Supplied were chairs of polished mahogany with blue covers and, in the tourist restaurant, furniture of natural sycamore, with upholstery in a floral patterned plasticized fabric. The tourist staterooms had furniture of light mahogany, blue upholstery and blue carpets. The quality of the panelling in the Scythia caused the greatest interest. It consisted of laminated blocks, two foot square and eight inches deep and consisting of 240 sheets of veneer. The blending of the veneers attracted further attention, being of sycamore, avodire, African mahogany, Makassar ebony, French and Australian walnut.

Three years after the death of Harry Morris the company was still attracting orders. There was the ending of utility furniture and new orders from shipyards. Neil Morris had designed a new furniture suite to be called, "Cumbrae." Originally the suite had been at the planning stage in December 1948 and introduced to the company in April 1949. Another two suites had been put into production and those were the "Pentland Suite" and the "Clan suite." For some reason the Pentland Suite was renamed, "Cumbrae". It was manufactured from natural oak and mahogany, becoming an overnight success.

What attracted the customer to the Cumbrae range was the quality of the workmanship and attention to detail of every piece. It was first displayed with a dining table and four upholstered chairs and a sideboard. Another attraction for the onlooker was the rich quality of the veneers, including Australian walnut, natural oak and mahogany. As a design technique, Neil Morris added an edging of pale sycamore wood. The cost to the customer of a Cumbrae Suite was 51 guineas free of tax. The company went onto a five-day week with orders of more than 100 for the Cumbrae suite. More work came in such as the decoration of the Royal Restaurant and the Rogano just off Royal Exchange Square in Glasgow. Cierva placed more orders for helicopter blades and Jacob Shapiro was delighted at the quality of the finish. Further orders were received for the Scottish Co-operative Wholesale Society and Grants furnishing with an unusual order for 100 golf club heads. Neil Morris was still designing interior work for private housing, such as houses at Pollokshaws and further work in Giffnock. Using techniques pioneered on the Mosquito aircraft, Neil Morris designed the Cloud Table. The Cloud was designed as an individual item, but was part of the Cumbrae range. It came in two sizes, a large size of four foot five in its widest part and a smaller version of three foot six inches. The top was made of six sheets of laminated wood, with a chamfered edge showing the six layers of Honduras mahogany, walnut and hackberry wood, with the top being mounted on four thick rectangular legs. The Cloud caused a sensation and, its odd shape proving highly influential in the post war years. Some Morris stockists brought out their own versions, but, in fact, they were debased versions of Cloud. Neil Morris had met Basil Spence at the Empire Exhibition in 1938 and the "Britain Can Make It" exhibition in 1946 and the "Enterprise Scotland" exhibition in 1947. In 1949 they collaborated on the "Allegro" range, which won a first class diploma in a Scottish furniture competition. Of the 114 entries, only eleven first class diplomas were awarded. The judges included Ernest Race and Geoffrey Dunn of Dunn's Bromley. Allegro was world beating and was exhibited with some of the finest contemporary furniture form Denmark, Sweden and America. Graham Sutherland designed the Morris exhibition, at the Kelvin Hall, Glasgow, in 1949. Neil Morris recognised that the Allegro range was the product of three men; Basil Spence, himself and Charlie Sim a long time employee of the firm. Allegro was worked in laminated Honduras mahogany and betula, inspired by the designs of Thomas Sheraton. The Allegro chair caused the greatest interest because of the method of manufacture. The chair was made from over 100 laminations, pressed together with synthetic resin and phenol formaldehyde, very similar to the construction techniques of the Mosquito aircraft using the Redux process. This technique made the chair incredibly strong and also allowed it to be machined down to size. But the arms, leg and back legs departed from the overall technique of the design, by using one lamination only. Manufacturing Allegro and the chairs was very expensive and it was only available to order. The chair attracted two special orders; one from the Victoria and Albert Museum and the other from the Museum of Modern Art in New York.

The Queen Mary returns to Southampton from her last trooping voyage

Morris also made ship nursery furniture

Queen Elizabeth on her first post-war voyage

The Queen Elizabeth first class lounge with Morris furniture and panelling

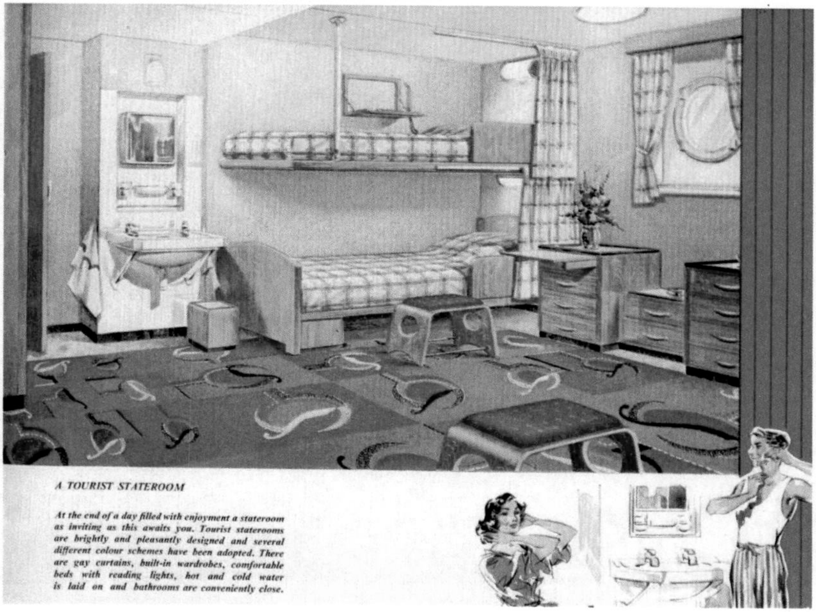

Ivernia Tourist Class

Laminates for panelling

Pendennis Castle

Ocean Monarch

Kelvin court refurbished 1947

Caronia, built at Clydebank, was the first major post war furnishing job for Cunard by Morris

The Caronia's cinema, Morris used their cinema seating experience in completion of this contract.

More work at Clydebank

The Carrick, formerly the City of Adelaide, at James Watt Dock Greenock awaiting movement to Glasgow Harbour.

The Empress of France was another post - war refurbishing contract

The Empress of Canada was used in the immigrant service to Canada as well as a passenger vessel.

Morris completed the tourist dining room of the Empress of Canada

The SS Scythia's first class smoking room, which caused a sensation by the use of a variety of panelling.

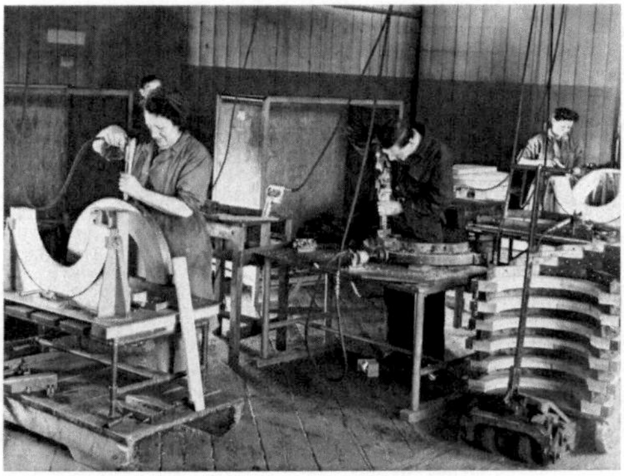

Completing tables by the use of machine tools, ensuring accuracy and a quality of finish. Long after the end of the war Morris still used female labour

Caronia chests ready for finishing. Note the timeless quality of the finish

Aquitania heads for the breakers, a fine vessel loved by all

CUMBRAE DINING SET edged with sycamore is made
in Australian walnut, natural oak and natural
mahogany. Shown here is the dining table, tea-
trolley, coffee-table, sideboard and armed-chairs.

The Cumbrae Range, which revolutionised the furniture market and swept away the concept of "Utility"

A Cumbrae writing bureau

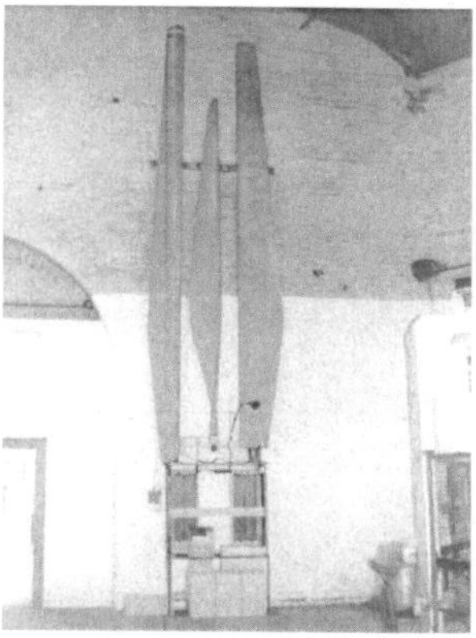

Cierva propellers found in storage at Milton Street

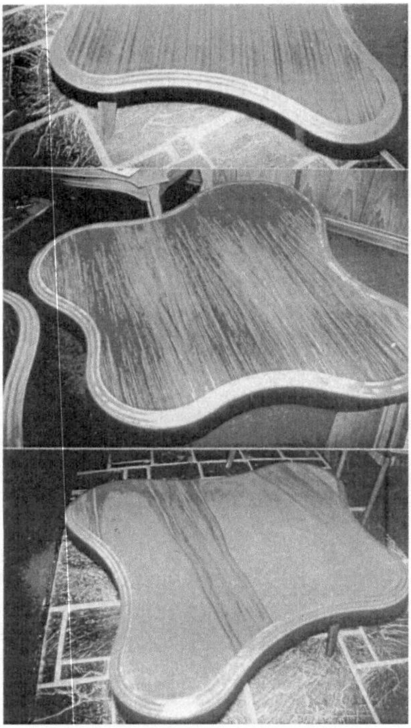

Prototype versions of the Cloud table seen at Drakemire, Glasgow

The Alexander Leyland Tiger bus, which was extensively panelled in walnut.

The Spence chair, part of Allegro, which owed more to the aircraft industry than traditional furniture practise.

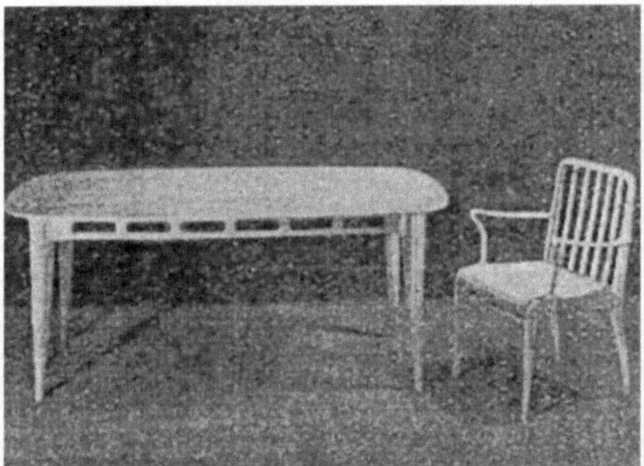

Allegro by Basil Spence and Neil Morris.

Building a Corporation of Glasgow double deck bus with Morris wood.

The later Empress of Britain, which Morris completed, including fixtures and fittings in addition to furnishing the cabins.

The Festival Ship Campania which was furnished for the 1951 Festival of Britain

The Arcadia for Cunard

Morris supplied panelling for the designs of Hugh Casson for the Festival of Britain.

Part of the Enterprise Scotland Suite, using seats with leather bands For the Festival of Britain 1951

Bambi stacking chair

The Hawker Hunter projected with Morris external fuel tanks

The Gift

Glasgow decided to present a wedding gift to the newly married royal couple, Princess Elizabeth and HRH Prince Philip, the Duke of Edinburgh. They had set up home in Clarence House and Glasgow's Lord Provost, Sir Hector McNeill, had appealed to the Glasgow public to contribute towards a gift. A committee was set up and it was decided to present the royal couple with the complete furnishing for Clarence House. Sycamore was chosen from the Balmoral Estate and shipped to Glasgow. Neil Morris despatched Jack Cantle to London to complete the initial measurements for the contract. Morris had been chosen for its reputation as a ship's furnisher and hotel outfitter. The fourteen pieces took seventy men more than nine months to complete. It was designed to be built into the sycamore-panelled rooms and all the lighting was concealed. The royal couple were involved in the design from the very start, but, at a later stage, when the pieces were being installed in Clarence House, Queen Mary was consulted and made a number of alterations to the final design. The Duke of Edinburgh had two wardrobes, one for his uniforms and the other for his civilian clothes and a cupboard for his shoes. Behind the wardrobe doors, were fitted mirrors. The fitted bookshelves were adjustable. Inside an alcove was placed the bed and, on each side, was a bedside table containing one shallow drawer over a cupboard. Beyond the small cupboards were built in open-ended bookcases and below these were fitted cupboards for shoes. More bookshelves were installed and underneath the bookshelves were drawers lined with African mahogany. In one of the walls between the windows was a dressing chest with eight drawers, and a recessed illuminated mirror. The fitted wardrobes were built into the walls, with double doors and inside full-length mirrors. All the interiors were of African mahogany and the handles were of silvered brass knobs. The People of Glasgow's donations ranged from sixpence to hundreds of pounds and they were invited to view the gift before it was shipped south to London. Clarence House has changed from the days of the late forties; H.R.H., the Queen Mother, took up residence and later, her grandson, Prince Charles. But one piece of furniture has followed Prince Philip and that is his writing table, which is now at Sandringham.

Clarence House furniture, made from Scottish Sycamore trees taken from Balmoral, Glasgow's 1947 gift to the royal couple. Designed by Neil Morris taking 70 men to complete under the direction of Jack Cantle. Bob McCafferty and Tom Partridge were instrumental in its manufacture.

Post War

It was just after the war that the company was hit with an excess war profit tax. It was the shape of things to come. The accountants were able to negotiate a deduction for fair wear and tear with the Commissioners of the Inland Revenue and the tax was reduced. During January 1950, Neil Morris called an extraordinary meeting of directors; he had a very important statement to make. The death of Harry Morris had revealed a great weakness in the company structure, which had been identified by the Board of Trade. It was a one-man show and not a limited liability company. As such, the company was liable for substantial death duties. This was to hit the company hard and it came from an unexpected source, the government. The government wanted its money and it wanted it now, no arguing. Neil Morris pointed out the gravity of the situation, in relation to orders received, work in hand and the production capacity of the company. Earlier, he had brought in a company of management consultants, who confirmed the findings of the Board of Trade. The company had no management structure to speak of and the wages the men and women were being paid at the time was not bringing out the full potential of the employees. Quality control consisted of the owner inspecting each piece of furniture as it was being completed and then passing for delivery, or literally throwing it across the yard if it was unsuitable. If anything happened to a machine, only the owner knew how to fix it. The viability of the company was now in question and whether to go on was the decision of Neil Morris. He explained further his ideas on what to do and closed the meeting. H. Morris & Co would still go on. He managed to negotiate the rate of return for Death Duties with the government and that would give him time. He opened new offices in London and Leeds and he floated a new company called, "Morris of Glasgow."

Orders were still coming in and the demand for the Cumbrae range was in excess of production. The Glenmill Laundry was purchased as a shadow factory for rifle production but the company left the helicopter and aviation field after the crash of the Cierva Air Horse. Hotel work was still coming in as were orders for fire doors for ships. A new contract was received from Barclay Curle to outfit a ship and install fire doors. But Neil Morris was uneasy with the prospect of more ship work. The boom in post war shipbuilding had continued and he had thought of setting up a ship-outfitting department at Milton Street. But, before he made the decision to actually set up the department he sent Jack Cantle around the British shipyards as a scout, to see if the idea was viable. Jack went to all the major and minor shipyards in Britain. His report was not encouraging. He pointed out that many of the finishing trades, including joiners and carpenters were continually being laid off and that the shipbuilding companies were very inefficient, lacking a sound cash flow: there were also the changes in aviation coming. Giant aircraft were being planned to fly the Atlantic. Wisely Neil Morris dropped the idea.

Panel and plywood work was still very much in demand. From Alexander's came an order for panels for Leyland Tiger single deck buses. Scottish Aviation started to manufacture buses on a sub-contract basis, with Morris receiving further orders for panelling and plywood. Plywood was also sent to Scottish Aviation for aircraft use, but after 1951 the company stopped manufacturing aviation plywood and left the SBAC. The Ministry of Supply gave the company some of its final defence orders. Work was received to make plywood masts for the Admiralty, in conjunction with Ferranti and guided missile work was received from the Ministry of Supply. Ship work came in from Vickers at Newcastle, John Browns and the other major shipyards the company had been working with since the early 1900s. For Vickers, they had three major merchant ships on the stocks requiring fitting out. They were 119, 122 and 123. Ship number 119 became the Ocean Monarch, Ship 122 became the City of York and ship 123 became the City of Durban. The cost of supplying 119 was £31176 for over 1300 pieces of furniture. For the new combined Union Castle Line, furniture and fittings was supplied for The Pendennis Castle, built by Harland and Wolff, Belfast, she was launched on 24 December 1957 following a delay caused by a dockyard strike. Consequently she was the last Union-Castle ship built by Harland and Wolff. She sailed on her maiden voyage to South Africa in January 1959, replacing the aged Arundel Castle. 763ft 2in long with a beam of 83ft 9in, her steam turbines gave her a service speed of 22.5 knots. With a GRT of 28 442 tons, she could carry 197 first class and 473 tourist class passengers. No new customers were to be taken on, Neil Morris having decided that ship furnishing was to wind down and not take up so much of the company's production capacity.

Festival and Exhibitions

Commemorating the Great Exhibition of 1851, the Festival of Britain was to include every city, town and community in Britain and it was the brainchild of Herbert Morrison. Tenders appeared in all the newspapers asking for companies to apply for work in all areas. H. Morris & Co was the number eight company on the festival list. They furnished the secretary's office and reconditioned the festival ship, H.M.S. Campania. As the "Campania," she was to sail around the coast of Britain exhibiting all sorts of British produce. The Morris detail was complete on this vessel, right down to the standing lamp in the captain's cabin.

The company also supplied panelling to many of the festival buildings. The panels were of veneers of Queensland walnut, desks were of solid oak with interiors of African mahogany and the secretary's desk was of gaboon plywood.

The Cumbrae range had been expanded, to include a cocktail cabinet, a coffee table, two dressing tables, a bureau and a lowboy. Four Enterprise Scotland dining chairs and fifteen tulip chairs were sent down to the exhibition in London. The tulip was a moulded plywood chair, manufactured with veneers of sycamore and mahogany. Sycamore was used on the veneer of the seat and backs, while mahogany was used on the facing of the laminated bent frame. These chairs were not to be used for purely display work, but were for use in the information bureau of the festival. The Cumbrae cocktail cabinet was displayed on a small stand in the Industrial Building of the South Bank. The idea was to display a small piece of high volume furniture, to show the high degree of craftsmanship and finish. As befitting its status, the Allegro chair was exhibited in its own right to a constant stream of visitors.

More Chairs

Bambi and Toby chairs are the most enduring designs of the company. Designed for stacking, the Toby combines metal and wood, with a six ply veneer of walnut. There is a suggestion of utility in its shape, through the spindly metal legs. Apart from the walnut version, there were another two; one with a painted surface and the other of a mahogany veneer with slightly thicker legs. For many years the Toby chair was a familiar sight in universities and schools or in cafeterias and was included in the Cumbrae range. The most important chair was the Bambi. It was moulded in one-piece plywood and dispensed with any screws, nails or joints. It was made from a six-ply sheet. Redundant sheeting was cut out and it was moulded and bent into shape. The only metal parts were the foot studs added after moulding, to give the chair a measure of steadiness on the floor. The redundant sheeting was also put to use when it was used to make trays - nothing was wasted. A later development of Bambi was upholstered and included in a composite design for the Institute of Contemporary Arts in London. Working with Neil Morris on this design were the architects Jane Drew and Maxwell Fry. Bambi is unique amongst one-piece furniture in that it is moulded in one piece, Neil Morris did it first.

Morestos was a material pioneered by the firm in its furniture. This was a synthetic glass fibre material, which was seen as having a great potential. Neil Morris saw that there was a use for it on aircraft, as well as in furniture production. He offered Hawker Aircraft a series of long range drop tanks for use in the Hawker Hunter. But the Hunter was in mass production for the Royal Air Force and NATO, which meant that the production line could not be interrupted and the Morestos fuel tank was dropped after tests.

Royalty

The Royal Family had used the battleship Vanguard in 1947 for a successful tour to South Africa. The consultant Mayfair specialist to the Royal Household was Miss Margaret Brigden and she had been consulted on the furnishings for Vanguard. Furniture from the Royal Yacht Victoria and Albert had been temporarily moved to Vanguard for the trip. But the royal yacht was far too old for the demands of a modern monarchy. The King had planned further trips to the Empire, particularly Australia, in 1951. Failing health meant that the king was unable to take on the planned voyage and Princess Elizabeth and Prince Philip agreed to go instead.

The Shaw Saville liner "Gothic" was taken in hand by Cammel Laird at Birkenhead and converted for the Australian trip. For her cabin work Morris supplied panelling of sycamore. Margaret Brigden had been responsible for setting up the team to refurbish Clarence House. Part of her team was Neil Morris. She attended all the Admiralty conferences and visited the shipyard to take measurements. She was to attend to the furnishings, curtains, carpets and bedspreads. But the subsequent death of the king meant that the expected voyage was cancelled and the cabin work on the Gothic was mothballed. After the coronation, the planned trip to Australia was revived again and work on converting Gothic was resumed at Birkenhead. Her funnel was raised seven feet by Thorneycroft to take the hot exhaust gasses away from the royal apartments and from disturbing the deck passengers. She also loaded some of the furniture and fittings from the old yacht Victoria and Albert. At Birkenhead a reporter from The Glasgow Herald had managed to get on board to view the furnishings of the substitute royal yacht: he described in clear detail what he saw, and found himself tip toeing through the cabins noiselessly. He added to his story "....A palace is a palace and not a fit object for idle curiosity." The press had heavily criticised the use of antique furniture for the voyage and they made it plain through their criticism that the Admiralty should have used the modern furniture that had been made for the Festival of Britain. Mysteriously Gothic called into the Clyde in November 1953 with very little publicity, docking at King George V Dock in the dead of night. There she took aboard further pieces of contemporary furniture, which were to be used in the Royal apartments. The plan was that the Gothic was to sail to the Caribbean and pick up the royal party, then sail through the Panama Canal to Australia. After the royal visit, the Gothic would meet the new royal yacht, Britannia, at Aden and there the fixtures and fittings from Gothic would be transferred.

The furnishing of the Royal Yacht Britannia was in the hands of Hugh Casson. Though John Browns at Clydebank had received the order it was Casson who was in charge of the furnishing committee. Hugh Casson had been involved in the organisation and planning of the Festival of Britain and many of the post war furniture displays. He was familiar with the work on Vanguard and Gothic and of Scottish furniture and furnishing manufacturers and was a friend of Neil Morris. A veil of secrecy was brought down on the Britannia, which still exists to this day. Many firms were indeed involved in her outfitting, but they cannot be traced. If they were involved with the outfitting the work that they did was seen as a donation. No fees were rendered or expected. However Morris supplied all the doors for Britannia and her panelling and much of the cabin furniture.

The expected changeover of furnishings did not take place at Aden, but at Tobruk in North Africa. (The early trials of Britannia revealed that she had engine problems.) Everything was taken out of Gothic and passed to Britannia; everyone took part in the changeover including the Queen and Prince Philip, Prince Charles and Princess Anne. They manhandled all the Victoria and Albert furniture across the bay and that included the Scottish furniture placed on board Gothic at King George V Dock

The Vanguard was used as the Royal Ship for the 1947 Royal Family trip to South Africa.

Fitting out the Gothic at King George V Dock Glasgow

The Gothic at the, "Tail o' the bank," awaiting the tide to take her to King George V Dock.

Furniture on the Gothic

Modern furniture on the Gothic. Much of it survives on the Britannia to this day.

Gothic in Australia

Britannia at Rothesay Dock Clydebank. Apart from Morris she was furnished by Wylie and Lochhead and Rowan and Boden

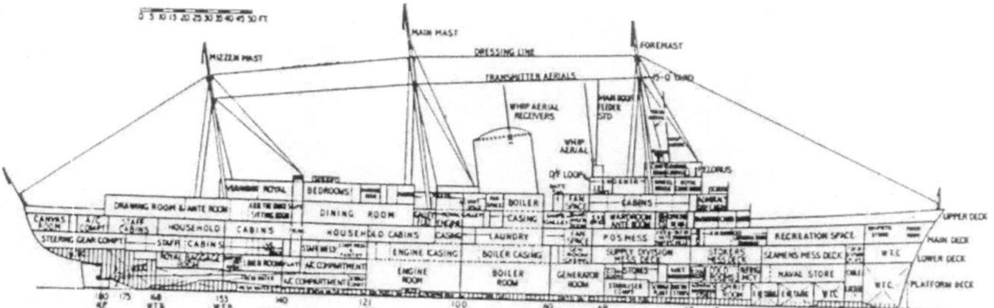

FIG. 11.—H.M. YACHT "BRITANNIA". PROFILE

The Royal Yacht Britannia

A Further Interlude

As a designer and interior furnisher the fame of Neil Morris swept across Europe. He was called in to design a display for the Gardner Lennox, New York, woollen showroom. They gave him nine months to do the work. He did it in two. Neil Morris saw the design in his mind as he had a telephone conversation with Gardner Lennox and, from the brief conversation, came a showcase, four chairs, a table and a cocktail cabinet. The most unusual piece was the glass-topped table, shaped like Loch Lomond to emphasise the union of the two firms - Gardners of Selkirk and Lennox of Loch Lomond. Engraved on the glass was a map of Loch Lomond while supporting it was a three-legged base of Australian Walnut. The four chairs were armless and of laminated birch with rubber-foamed seats covered in grey brown blue and red material. The cocktail cabinet was of walnut and ash lined with sycamore. The doors had inset twisted copper wire underneath the veneer, flattened at high pressure to give a faint green marking. The two - part showcase was made of walnut and ash for cashmere and tweeds. The top was divided into sycamore lined sections kept dust proof by glass sliding doors. There were four deep drawers for samples of wool and fabrics and above those were sliding trays set into side grooves. The design caused a sensation in New York even though it was designed to boost the export demand for textiles.

Neil Morris was asked to transform room 266 at the Caledonian Hotel in Edinburgh. The idea was to make the room usable during daytime and not just be a bedroom. The contract took six weeks to complete. It became a bed sitting room in the day and an elegant bedroom by night. He installed a writing table of Honduras mahogany and two divans, which became beds in a matter of minutes. Beside the beds he placed a coffee table and, between the beds, fitments to take the bed linen. The tops of the coffee tables were hinged to form luggage stools; twin wardrobes were at the end of the room and between them was a sideboard. One wardrobe was fitted with hanging and shelf space while the other camouflaged a toilet basin. The sideboard gave additional storage space. The Honduras mahogany had been specially chosen by Neil Morris to withstand the rigours of central heating. Neil Morris was quoted as saying to the press, "....I have always advocated dual purpose furniture as the answer to the space problem in hotels, as well as in homes. Many people today have to live in one room, which by using dual purpose furniture can be quickly switched from bedroom to sitting -room."

For Suvretta House, in Saint Moritz, Neil Morris adapted Cumbrae for Mr Candrian - Bon. Mr Candrian - Bon had come to Glasgow to ask his advice on how the house was to be outfitted. The design once again attracted the notice of the press and the work was well received.

A Fresh Chapter

Due to the in - depth nature of the work now being undertaken at H. Morris & Co., Neil Morris took a three-month break from his duties. Since the death of his father in 1946 he had been striving to maintain the independence of the company and had been continually involved in all the company operations, never taking a break. As a married man he had his family to consider. While he was on leave there came a bid for the company from a firm called Camp Bird but the bid was repulsed. On his return Neil Morris obtained overseas contracts for Cumbrae from as far afield as South Africa and Australia. Hotel work was also received from the Central Grand Hotel in Khartoum, Sudan and there was even a plan to set up a hotel in Nairobi, Kenya. The Lorne Hotel, in Glasgow was also surveyed with a view to purchase but nothing came of the project. From the United States Air Force (USAF) in West Germany, Morris received the contract to outfit all the cafeterias and officer's clubs at airfields throughout the country.

In 1960 Neil Morris flew to the Soviet Union. The visit there had been cleared by the British and Soviet authorities so that Neil could travel in freedom all round all the Baltic States including the old state of Petersburg, which was dominated by the city of Leningrad. After the collapse of the Soviet Union the city reverted to its old name, "Saint Petersburg." The extent of cooperation between the British and Russians was such that Neil Morris was allowed to take his camera and take as many pictures as he wanted without any security supervision. The visit was to last a month.

The next year he visited Germany as a follow on for the supply of furniture to the USAF. While in Berlin he dined with his old friend Hanna Reitsch, the aviatrix, renewing his friendship from pre - war years. During the war years she had served as a test pilot for the Luftwaffe, flying helicopters, gliders and rocket planes such as the ME 163 and in doing so had broken her back when she flew a piloted example of the V1 flying bomb. From Berlin, Neil travelled to Cologne for a furniture fair, and then he went into France and completed his trip at Nice. He oversaw work in the Belgian Congo for furniture for a colonial hospital. The panelling was in Formica, but there are no details of how the work was to have been completed. The work undertaken in Germany for the USAF bore further fruit, with cabinetwork being completed in the American Embassy at Grosvenor Square, London. Jack Cantle followed up this order with more contracts for filing racks in the embassy offices.

Then the drawing office came up with a new design called Bianco. This design departed from the usual Morris practise of using natural woods and was manufactured with chipboard and a variety of veneers in its finish. Thermal bonding was also introduced for modern bureau units and more orders came in from Grosvenor House (This type of bonding work was pioneered in the construction of the Mosquito.)

From Europe came Slovakian oak shipped in the SS Castel Bianco, a Victory Class freighter and from Messrs Brownlee came supplies of red oak for manufacturing. For Kirkle Drive in Whitecrook, Glasgow, a private house, Neil Morris designed the interior furnishings and panel work, the furniture and panelling coming from Milton Street. Some prototype work came in from Rowan and Boden for the design of chairs and from Cunard an order was received for a folding chair called a "Sundowner." From 1923 their existed one piece of furniture that had been in continuous production, except for a wartime break and that was the writing table. Termed design 875 in the product line book the writing table had reached the end of its usefulness. People no longer wrote in the manner of their fathers preferring the telephone to the ink pen. There were also the ravages of temperature in the central heating systems then being installed in most homes, but above all it was too expensive to manufacture. In 1963 the last one was completed.

Clydeside was also changing. The shipyards were reeling under the lack of orders due to the popularity of flying. It was now cheaper to take an aircraft than to sail in a ship. Cunard ordered their last liner from John Brown at Clydebank and this vessel was to be named Queen Elizabeth II. The original Queen Elizabeth had been sold off as a floating university but had been destroyed in a fire in Hong Kong harbour. The Queen Mary had also been bought by the Americans and was now a floating hotel/museum on the west coast of America. For some reason Cunard felt that there was a demand for a new passenger liner and the government had promised substantial funding for the ship.

H. Morris & Co received the order to panel the vessel and supply all the fixtures and fittings for the First Class Cabins and all the doors for the vessel. Bedside cabinets became the responsibility of the youngest apprentice at Milton Street - Robert Morris. The captain's cabin was also Morris outfitted, as well as the doctor's cabin. With the introduction of metrication the old days were fading fast, along with pounds, shillings and pence. The Clyde shipyards were merged by the government into a single group, Upper Clyde Shipbuilders but reflecting the direction of the times the group collapsed in August 1971. (Today there are only two shipyards in operation in the upper Clyde, but they concentrate on military orders. Surprisingly an aircraft manufacturing company now owns these yards.)

H. Morris & Co., Ltd. received further orders from ICI and the new Strathclyde Regional Council (SRC) Unitarian authority. The regional council was an amalgamation of all the West of Scotland town councils and the old town councils or "Corporations" became district councils. One order for SRC was to furnish the chief executive's office in India Street, Glasgow. Orders also came in to furnish offices and churches and the new fast food restaurants, which were appearing all over the country. Exhibitions of Morris furniture still took place at Glasgow and Edinburgh and there was still a huge demand for quality furniture. In 1976 Neil Morris retired and Robert Morris became managing director, the third member of the family to do so. When he entered his father's old office what he first noticed was the size of the desk, which was enormous. That same year, NATO placed a huge furniture contract and an order was received to refurbish the American Embassy at Grosvenor Square, London.

Britain can make it

The Loch Lomond table for the New York display

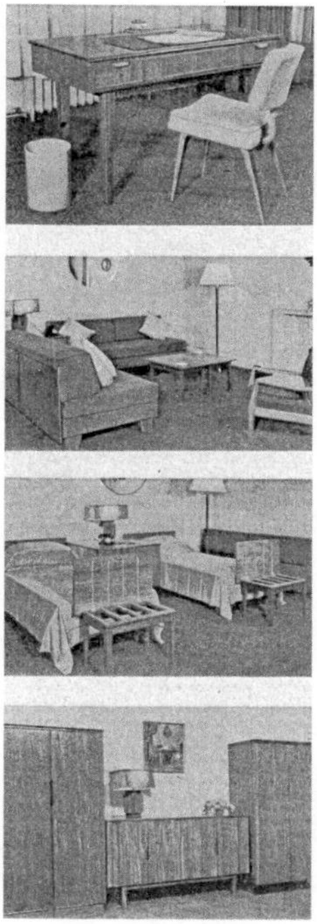

Furnishing room 266 in the Caledonian Hotel Edinburgh.

Processing wood, once again by female labour for the Cumbrae range.

Hanna Reitsch, pioneer pilot

The Q E 2 on trials on the Clyde.

Queen Elizabeth, after being sold by Cunard, to be converted in Hong Kong as a floating university during 1971.
A few weeks later she caught fire and was a total loss

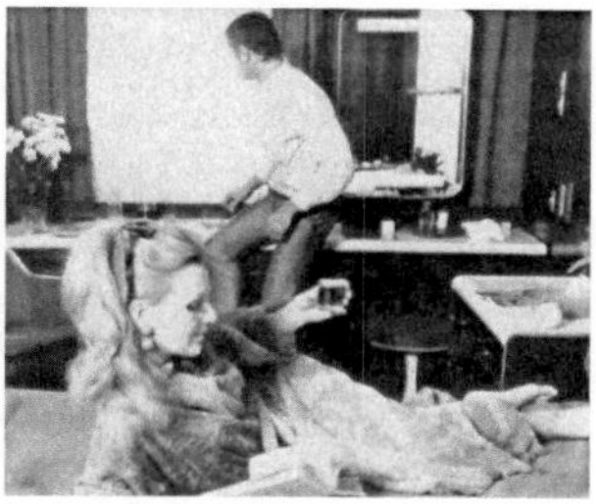

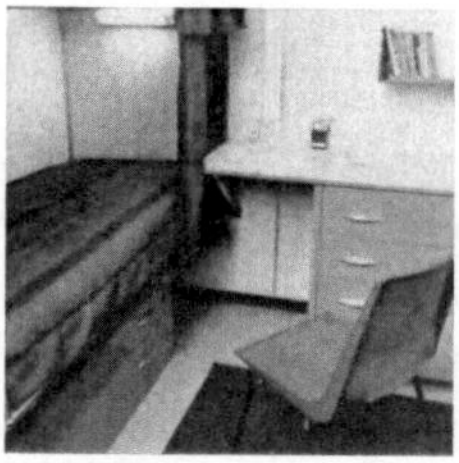

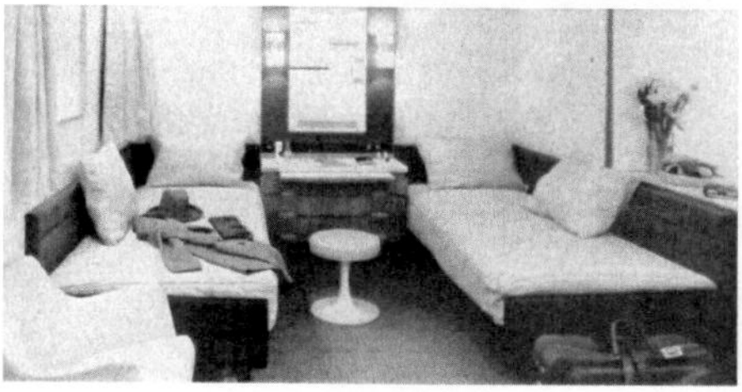

Cabin work on the QE2. Cunard demanded simplicity in the cabins with a touch of luxury

Rooftops across Milton Street

Outside the old building at Milton Street

Tait's Tower 1938 Empire Exhibition

Morris furnished the restaurant

Empire Frome refurbished by Morris with general cabin work including decking

Neil B. Morris